TRANSPORT LESSONS FROM THE FUEL TAX PROTESTS OF 2OOO

To our wives Joanna and Alison for their enduring patience during a year of heightened research workload triggered by a few days of distruption to the UK transport system in September 2000.

Transport Lessons from the Fuel Tax Protests of 2000

Glenn Lyons and Kiron Chatterjee
Transportation Research Group, University of Southampton

Ashgate

Published by
Ashgate Publishing Limited
Gower House
Croft Road
Aldershot
Hampshire GU11 3HR
England

Ashgate Publishing Company
131 Main Street
Burlington, VT 05401-5600 USA

Ashgate website: http://www.ashgate.com

British Library Cataloguing in Publication Data
Transport lessons from the fuel tax protests of 2000. -
(Transport and society)
1. Fuel -Taxation - Great Britain 2.Protest movements - Great Britain 3. Motor vehicle driving - Great Britain
I. Lyons, Glenn II. Chatterjee, Kiron
336.2 ' 78 ' 6655 ' 0941

Library of Congress Control Number 2001098287

ISBN 0 7546 1844 7 ✓

Printed and bound in Great Britain by Antony Rowe Ltd.,
Chippenham, Wiltshire

Contents

List of Contributors

Colin Bamford
Transport and Logistics
The University of Huddersfield
Huddersfield HD1 3DH
UK

Susan Beatty
Drug Misuse Research Unit
School of Epidemiology and Health Sciences
University of Manchester
Kenyon House
Mental Health Services of Salford NHS Trust
Bury New Road
Manchester M25 3BL
UK

Mark Beecroft
Transportation Research Group
Department of Civil and Environmental Engineering
University of Southampton
Highfield
Southampton SO17 1BJ
UK

Michael Bell
Transport Operations Research Group
Claremont Tower
University of Newcastle
Newcastle upon Tyne NE1 7RU
UK

Peter Bonsall
Institute for Transport Studies
University of Leeds
Leeds LS2 9JT
UK

Paul Boulter
TRL Limited
Old Wokingham Road
Crowthorne
Berkshire RG45 6AU
UK

Kiron Chatterjee
Transportation Research Group
Department of Civil and Environmental Engineering
University of Southampton
Highfield
Southampton SO17 1BJ
UK

Michael Coyle
Transport and Logistics
The University of Huddersfield
Huddersfield HD1 3DH
UK

David Eves
TRL Limited
Old Wokingham Road
Crowthorne
Berkshire RG45 6AU
UK

Phil Goodwin
ESRC Transport Studies Unit
Centre for Transport Studies
University College London
Gower Street
London WC1E 6BT
UK

Reg Harman
2 Valley Close
Hertford SG13 8BD
UK

John Hickman
TRL Limited
Old Wokingham Road
Crowthorne
Berkshire RG45 6AU
UK

Glenn Lyons
Transportation Research Group
Department of Civil and Environmental Engineering
University of Southampton
Highfield
Southampton SO17 1BJ
UK

Greg Marsden
Transportation Research Group
Department of Civil and Environmental Engineering
University of Southampton
Highfield
Southampton SO17 1BJ
UK

Michelle Meadows
Psychology Division
School of Sciences
Staffordshire University
College Road
Stoke-on-Trent ST4 2DE
UK

Robert B. Noland
Centre for Transport Studies
Department of Civil and Environmental Engineering
Imperial College of Science, Technology and Medicine
Exhibition Road
London SW7 2BU
UK

Graham Parkhurst
ESRC Transport Studies Unit
Centre for Transport Studies
University College London
Gower Street
London WC1E 6BT
UK

John Polak
Centre for Transport Studies
Department of Civil and Environmental Engineering
Imperial College of Science, Technology and Medicine
Exhibition Road
London SW7 2BU
UK

James Quick
TRL Limited
Old Wokingham Road
Crowthorne
Berkshire RG45 6AU
UK

Paul Rosen
Science and Technology Studies Unit
Department of Sociology
University of York
Heslington
York YO10 5DD
UK

Patrice Salini
Institut Management transport Logistique
Université Paris 12
Avenue Pierre Point
77127 Lieusaint
France

Neil Thorpe
Transport Operations Research Group
Claremont Tower
University of Newcastle
Newcastle upon Tyne NE1 7RU
UK

David White
Psychology Division
School of Sciences
Staffordshire University
College Road
Stoke-on-Trent ST4 2DE
UK

Tony Whiteing
Transport and Logistics
The University of Huddersfield
Huddersfield HD1 3DH
UK

Foreword

For a brief period in the autumn of 2000 the UK and several other European countries, including France, Germany, Belgium and the Netherlands, faced disruption in the face of widespread protests against high petrol and diesel prices. The episode was referred to as a fuel crisis. Many commercial businesses heavily dependent on fuel felt under extreme pressure as rising prices squeezed their already slim profit margins. An initial series of blockades in France by fishermen, lorry drivers, farmers and taxi drivers won concessions from the French government. This provided a catalyst for protest to take place in other neighbouring countries. The protests highlighted the extent of Europe's dependence on fossilised fuel for mobility and the sensitivity of the public to the cost of motoring at a time when governments are anxiously trying to discourage car use in the face of growing road traffic congestion.

In the UK the fuel crisis (at least as far as the media would have us believe) brought the country to the brink of paralysis as fuel supply was constricted and widespread panic buying at fuel station forecourts ensued. As a result a nation of car users faced a number of days during which fuel had seemingly become a finite resource demanding that people reappraise their daily routines of activity, travel and use of the car.

The fuel crisis was and remains politically very sensitive. It raised issues for economics and business processes. Perhaps most poignantly it created a temporary environment rich in potential for the study of travel behaviour response and adaptation. In short the crisis, aside from the immediate disruption it caused, provided an opportunity for fresh insights into some key transport issues - an opportunity for lessons to be learnt.

This book harnesses a collection of works that endeavour to explore the nature, consequences and implications of the UK fuel crisis. We hope it will prove to be one of the principal sources of reference concerning the crisis in terms of its significance to the transport profession. At the time of writing we are aware of other works that should act as a complement to this volume (many of these are addressed in Chapter 12) and there will no doubt be further publications in due course. From our own study of the crisis we are now in possession of an extensive database of information on car users' experiences and we ourselves would anticipate wishing to explore the topic further, building upon the key findings we report in this book.

The fuel crisis of September 2000 was perhaps notable for the important role played by information and communications technology. There have been previous episodes of fuel disruption but this was the first occasion where mobile phones and the Internet were used, and used with great effect, by protesters. The media played their full part in delivering the unfolding events and details of the crisis directly into our living rooms. Researchers too have found the web and email invaluable aids to their endeavours.

In considering the fuel crisis there is the inclination to contrast it with normal conditions and circumstances. Yet it would seem that 'normal' is becoming itself increasingly difficult to define. In our modern world we face a continual turnover of new national and global events and dilemmas that are shaping attitudes, behaviours and constraints. Perhaps one overarching lesson therefore from the fuel crisis is the need to intensify our efforts to understand travel behaviour since such understanding will be at the heart of forging effective transport policies that are not simply reactive to events.

There is no doubt that the emotive issues at the heart of the fuel tax protests in the UK and elsewhere in Europe will remain a topic of intense interest, deliberation and even further protest. On 12 September 2001 the European Commission published its Transport White Paper 'European transport policy for 2010: time to decide'. In this it highlights a need to harmonise fuel taxes across Europe to improve the effective functioning of the internal market. Such harmonisation would almost certainly see a call for the UK (which has the highest levels of fuel tax in Europe) to reduce taxation on fuel.

Who can say when we might face a further fuel crisis? Exactly one year on from the UK fuel crisis, the use of the word 'crisis' was set into context by the horrific terrorist attacks in New York and Washington. With the possibility of further instability in the Middle East we should not be too quick in forgetting to what the fuel crisis exposed us. Oil, and in turn fuel supply, is the very lifeblood of the functioning of our modern societies - we must be mindful of the dire consequences that will face us should this lifeblood cease to flow.

A Reminder of the Events of September 2000 in the UK

The crisis in the UK was short-lived. At the time of this book's preparation many people will have still been familiar with the period of time concerned. However, as time passes memories will fade. It is important therefore at the book's outset to provide a brief record of events that constituted the UK fuel crisis. The BBC documentary series 'Panorama' usefully broadcasted a programme shortly after the crisis entitled 'Pump Wars - 7 days that shook the Government' which provided a day-by-day portrayal of the crisis. The account below is based on this programme.

Day 1 - Thursday 7 September

On Thursday evening a convoy of local farmers and hauliers arrived at the Stanlow oil refinery in Cheshire to protest at fuel prices 'headed' by a local farmer - Brynle Williams. The aim of the protest was to stop tankers leaving the refinery - the protesters were seeking the same effect as in France where blockades of oil refineries and roads had forced the French Government to listen. All of this was against a backdrop of a world price of crude oil that had nearly tripled since the Labour Government came to power in 1997.

Day 2 - Friday 8 September

The protesters opted to stay within the law and maintained a peaceful protest. Nevertheless, tankers inside the refinery were now not moving - drivers were not prepared to pass through the gates and effectively petrol supplies from Stanlow were 'choked off'. Meanwhile the protest was spreading elsewhere. By Friday night the Pembroke refinery was 'shut down' along with other refineries 'across half of Britain' yet with, it seemed, 'barely a whisper of public protest'. Transport Minister Lord MacDonald dismissed the notion of direct action as 'not the British way at all'. It was apparent in hindsight that the Government failed to recognise what was going on and to prioritise it. When looking back on the crisis Andrew Smith MP, Chief Secretary to the Treasury, acknowledged that 'it was a number of days before the scale of the situation became apparent'.

Day 3 - Saturday 9 September

In France the blockades were being lifted following agreements reached between the protesters and Government on cutting the cost of diesel fuel. Meanwhile queues were forming at British filling stations and supplies began to run down at an alarming rate. Scotland was the first to run out of fuel after the Grangemouth refinery 'closed down' - the effects were being felt throughout the North and soon garage forecourts were being closed.

Day 4 - Sunday 10 September

A farmer in the South West, seeing the success in France where the Government had been forced to listen in a matter of days, contacted the local media indicating that protesters would shut down the refinery in Plymouth on Monday evening. This was an apparently calculated and effective move to provoke panic buying and thanks to the media the public did the rest in emptying the South West of fuel faster than anywhere else in the country. Meanwhile it seems the Prime Minister's attention remained occupied by the successful bid to secure release of hostages in Sierra Leone. By late Sunday evening the Government had still not responded to the oil companies calls of concern.

Day 5 - Monday 11 September

Despite the presence of police, drivers responsible for carrying tankers of explosive fuel were still concerned about their safety and refusing to transport fuel. There was a suggestion that some drivers were genuinely concerned about their safety, whilst others seized upon it as an excuse. The Prime Minister became more acutely aware of the protest situation and oil companies' minutes record a lot of Government pressure to get the petrol moving. Meanwhile, by midnight, closure of the Purfleet refinery in Essex had sparked panic buying in the South East - by nightfall 4000 filling stations across the country had run dry.

Day 6 - Tuesday 12 September

A report from the Government's emergency management committee Cobra (assembled only in times of national emergency) warned that the situation was near breaking point and that food shortages could be expected within a few days. The Prime Minister could not understand why the oil companies could not get their tankers moving when at Stanlow the road was clear and there was no obvious intimidation. Ministers were astounded to learn that Shell tanker drivers were employed by someone else - the haulage contractor P&O. Bill Morris, General Secretary of the Transport and General Workers' Union, observed that it is difficult to instruct someone if that person is not employed directly. He suggested that this represented 'one of the lessons that had to be learned out of this dispute' - policing and national security through the army is not subcontracted out so why should private companies be permitted to subcontract out a vital national resource? P&O remained silent though as a haulier it would benefit from any reductions in fuel price that the protests might bring. By this stage in the protest the protesters were those deciding who should qualify for emergency fuel. That evening the Prime Minister held a press meeting and signalled that he was in no mood for compromise. The protesters began to realise that the scale of the protest was becoming unmanageable and that they were now running a risk of losing the public support they had been increasingly enjoying. They resolved to maintain their protest for another 24 hours.

Day 7 - Wednesday 13 September

The fuel shortage was now very visibly biting across the country but further pressure was added in protest terms with lorries blockading Park Lane in the heart of London. The Government launched a two-pronged attack. The first was to stress to the nation how the protest was affecting the National Health Service. In a television broadcast from the Prime Minister he commented that 'lives are at risk if these people cannot get to work'. Meanwhile the Transport and General Workers Union became involved in negotiations at Grangemouth in Scotland to secure the support and cooperation of drivers in the hope of achieving a domino effect elsewhere. Provision of police escorts where needed and a doubling up of tanker crews persuaded the drivers to begin once again transporting fuel. The blueprint devised for Grangemouth was then quickly adopted at all refineries.

Day 8 - Thursday 14 September

At Stanlow oil refinery in the early hours of Thursday morning the protesters decided to end their protest whilst it was still on their own terms - they were concerned about the increasing risk of blame for a hospital death being placed at their feet. The protest was stopped but a 60 day ultimatum left with the Government to reduce fuel taxes - this was timed to coincide with the Chancellor's November budget statement. After the crisis Brynle Williams stated that 'what we want is parity with Europe'.

Contents of the Book

An Appreciation of the Roots of the Fuel Tax Protests

The book is divided into three sections. The first of these explores some of the underlying issues that precipitated the fuel tax protests.

Chapter 1 reminds us that the fuel tax protests of 2000 were not the first events of their kind. During the 20th Century there were four occasions when shortages in petrol have had a significant impact on the UK and other countries. Harman draws together the details of events and circumstances on these occasions which enables some interesting parallels and distinctions to be drawn when compared with the events of 2000. He notes that importance of the car has increased dramatically over the period considered but, despite this, some lessons apply equally today. One obvious lesson is the urgency to reduce reliance on imported transport fuel.

In Chapter 2, Rosen also uses past events to illuminate the present. He looks at how political pressure and activist groups have influenced the direction of transport policy in the years leading up to and including the fuel crisis. The objectives and methods of protest of the anti-car and pro-car sides of the argument are compared. Rosen contrasts the broader aims of environmental protest to move towards a potentially more sustainable long term future with the seemingly more narrow personal interests exhibited by protesters and the wider public during the fuel crisis. Yet on both sides of the debate protesters are learning from one another and the ability to orchestrate effective protest making use of modern communications technology is highlighted. The conclusion is drawn that it is difficult to predict what will happen in the future as positions shift in ways that are strategic, not entirely consistent and are not only related to transport issues. Rosen argues that periods of activism based on arguments of individual freedom may have limited life spans as they tend to result in the problems for society of their ambitions becoming evident.

Chapter 3 focuses on the country that provided the catalyst for the UK fuel tax protests. In France 'traffic embolism' has been a longstanding means used by those wishing to put pressure on the government. Salini explores the factors that gave rise to the French fuel tax protests and discusses the roles and standpoints of the various trade unions and employers' associations. After many years as a protected and regulated national sector, the French haulage industry has been exposed to European integration and regulatory reforms with common rules based upon non-discrimination and the free market. This has forced the industry, with a mix of large and small players, to massively improve its productivity, set against rising road transport prices.

Chapter 4 examines a significant Government policy which, alongside changing world oil price, has been principally responsible for the dramatic increases in fuel prices on filling station forecourts in the UK. The fuel tax escalator, which has now been abandoned, was a measure to increase fuel prices in real terms every year. In this chapter, Marsden shows how environmental reasons have been prominent in Government justification of the escalator after its introduction in 1993 but were conspicuous by their absence in the Government's defence of fuel tax levels during the fuel crisis. An assessment is made of the effectiveness of fuel taxes in making motorists pay for social costs including environmental externalities. Marsden shows that congestion costs dominate total external costs and that it is impossible to reflect environmental or congestion externalities in a tax based on the consumption of fuel. Furthermore, relative rates of taxation on petrol and diesel have not been set to reflect their environmental impacts. Marsden concludes that although the fuel duty escalator has helped deliver reductions in CO_2 emissions the Government's current taxation policy on vehicles is best described as a tax with environmental linkages rather than an environmental tax.

Chapter 5 raises a thought provoking question, namely - are fuel prices important? Goodwin notes firstly that the protests of 2000, in spite of being obviously labelled as fuel tax protests, were likely to have been the culmination of public unrest concerning a number of policy issues both within and outside transport. Nevertheless, public feeling over fuel prices was undoubtedly a key stimulus for the protests. This chapter takes a look at sensitivities associated with fuel prices. The assumed degree of sensitivity of car use to fuel price (referred to as a price elasticity) has a significant influence on road traffic forecasts (and by implication transport policy). The chapter highlights the discrepancy that has existed between values of elasticity assumed by Government and those deduced from a large body of empirical work. It emphasises the importance of using values based on empirical research even when they are quite close to zero. The question is posed as to what low price elasticities mean. Low elasticity implies on the one hand growing public indifference to price with the temptation to therefore also attach growing political indifference to it. On the other hand it implies car use is a necessity rather than a luxury and therefore there is an inability for the public to adjust behaviour to price change - this suggests that price increases will meet strong public resistance. Goodwin concludes that fuel prices remain important influences on both travel behaviour and political behaviour and improved understanding of price sensitivities will be very helpful to future transport decision making.

Empirical Evidence of the Impacts of a Fuel Shortage

The second section of the book provides various pieces of empirical work concerning the UK fuel crisis.

Chapter 6 presents data concerning traffic conditions on three of the UK's motorways - the M25 around London, the M6 near Birmingham and the M27 near Southampton. The loop detector data covers periods both before and during the fuel crisis allowing a comparison of daily traffic flows, daily flow profiles, daily speed profiles, speed/flow relationships, vehicle composition, estimated journey times and emissions levels. Eves et al. show that substantial reductions in motorway traffic flow levels occurred during the fuel crisis with reductions in flow generally of less magnitude in the peak periods. The lower flows meant that the usual flow breakdown in the peak periods did not occur at any of the locations and therefore journey times across the day were higher and more consistent. The findings also appear to support the popular perception at the time of the crisis that people were driving more slowly to conserve fuel. Emissions calculations suggest that emissions were reduced in line with traffic reductions, with lower vehicle speeds making little impact.

In the seventh chapter, Chatterjee and Lyons report on a mailback questionnaire survey of car users conducted immediately after the crisis to determine how people reacted to the circumstances faced during the crisis in terms of their daily routines and travel choices. Detailed accounts were received from over 1600 individuals. The chapter presents and discusses many detailed findings from the survey, highlighting the degree of use of, and satisfaction in using alternatives to the car. The survey showed that about one-fifth of those who use the car for commuting and for school escort journeys changed mode and that a substantial amount of business and leisure travel was not undertaken. Implications of the survey results for policies to reduce car use are drawn out. It is noted that events like the fuel crisis force people to re-examine their lifestyle and travel decisions and can trigger long term changes to these. However, great care must be taken in assessing whether responses during the fuel crisis are likely to be indicative of those that would be chosen as a longer term adaptation to policy interventions.

Chapter 8 also reports on a survey of car users. The survey took the form of telephone interviews of 1000 car users undertaken two months after the crisis asking them about their responses during the crisis and what they would do if another fuel crisis occurred. Thorpe et al. seek to identify from the survey the prospects for significant shifts in behaviour and the policies that could prompt them. They found that working from home is only an option for some people for some days and that public transport, walking and cycling are considered to be more realistic alternatives to the car. Nevertheless, the work suggests that bus travel must be reinvented if it to offer a viable alternative to the car and that greater attention should be paid to demand responsive services. Enabling people to better plan their travel and activities is also highlighted as a means of influencing car use.

Chapter 9 looks at car user behaviour from a psychological perspective. Beatty et al. use a five stage framework (the transtheoretical model) of behavioural change in relation to car users' willingness to reduce car use. Individuals are associated with and can move forwards (or backwards) between the five stages of precontemplation (not yet thinking about change), contemplation, preparation, action and maintenance (when all possible change has been effected). In a study of 855 respondents in April 2001 the proportion of the sample in each stage of change was identified. After the fuel crisis a sub-sample was approached again. The survey results show some random movement in stage positions but a net movement to later stages. Whilst the authors acknowledge some limitations in the applicability of the transtheoretical model to car dependence, they suggest that preparers would be a worthwhile group to target in terms of reducing car dependence. They were more receptive and most likely to change stage and were more likely to say that they had changed their travelling behaviour in lasting ways and changed their opinions during the fuel crisis. Car use reduction strategies to target those at different stages are suggested based on the study results.

Chapter 10, in contrast to the preceding chapters in this section, moves to the issue of road haulage and the significance of fuel price and availability. This chapter highlights the pressures faced by an industry operating with slender profit margins. Whiteing et al. set out the role of fuel costs for freight transport operators and describe the results of a survey of individual operators on their fuel management strategies during the fuel crisis, in the 60 day period after the fuel crisis and beyond that period. Some 50 operators of commercial vehicle fleets participated in the survey. The survey revealed that many fleet operators set short term contingency plans in place immediately following the fuel crisis to maximise the availability of fuel. In spite of short-term and generally short-lived changes to fuel management, the findings reveal that the fuel crisis has done little to trigger substantial lasting changes to operations, suggesting that the road haulage industry has already attained a high level of effectiveness in its fuel management with little room for further improvement.

Learning from Crises

The third and final section of the book reflects on what can be learned from the fuel crisis and indeed crises in general.

Chapter 11 is a companion chapter to Chapter 7, concerning the survey of car users that was conducted immediately following the crisis. The survey might be considered to have been a research crisis to follow the fuel crisis. Both lasted only a few days. Lyons and Beecroft present and discuss the experiences of conducting the survey. The process followed was not altogether untypical of a personal travel survey. However, the defining characteristic of this particular study was the time frame. On Thursday 14 September a decision to conduct a survey was made. By Friday 22nd September 5000 mailback questionnaires had been distributed in Southern England with a further 6000 questionnaires being distributed elsewhere. The obstacles, opportunities and challenges faced in pursuing the survey are highlighted with reflection on issues that contributed to its eventual success and those that served to compromise the survey's achievement. Lessons on planning for future emergency surveys are identified.

The fuel crisis arose almost unannounced. It has been likened to an experiment and earlier chapters have sought to understand how participants behaved within the experiment. However, as an unplanned event there was no coordinated effort to study its consequences. Both during the crisis and in its wake numerous sources in the press and within organisations affected by the crisis or with an interest in its consequences have provided their individual commentaries and figures. Chapter 12 has sought to harness the diversity of information surrounding the fuel crisis and to draw from this any further observations and lessons that complement those provided elsewhere in this volume. Marsden and Beecroft present information obtained from a range of sources covering lifestyles, media, business, telecommunications and transport. There were winners and losers in retail with local food stores and supermarkets reporting increased sales but department stores and out-of-town shopping centres reporting reduced sales. There were substantial increases in patronage on light rail and rail but only small increases on buses. Since the data upon which the chapter is based were not always intended to provide insight into the consequences of the fuel crisis, the chapter raises some unanswered questions. However, it concludes that the events of September 2000 reflect a drama rather than a crisis – the media enjoyed the opportunity of making the most of the problems that arose, but in fact many people were not badly affected and there were winners as well as losers.

Chapter 13 looks at the implications of the fuel tax protests for future motoring taxation policy. Parkhurst notes that although the Government failed to make large cuts in fuel tax at the time of the crisis, it promised not to make future increases in fuel duty. The Government will also not be able to increase taxation on new fuel types as it is committed to encouraging their take up. Taking these constraints into account, the impacts on the tax revenue received by the UK Government are calculated for different scenarios of traffic growth in the next 10-15 years and advances in the fuel efficiency of the motoring fleet. Tax revenue will decline under all scenarios. It is determined that lower income groups will find motoring less affordable as they will not benefit from the more efficient vehicles that are available. As a result of these findings, Parkhurst suggests that there will be a more positive environment opening up for the introduction of wide-scale road pricing.

Chapter 14 looks critically at what can be learnt from transport crises by taking the fuel crisis as an example. Bonsall assesses the wider applicability of the findings described in the rest of the book and in doing so provides a fitting conclusion to the book. He asks what might be learned to assist in the management of future crises or in the pursuit of transport policy objectives. More generally, the chapter considers whether the behaviour of individuals, groups and systems during a crisis can help us to understand or predict behaviour in other situations. Crises involve a number of defining characteristics that both render them different from one another and distinctly different from 'normal' circumstances. It is noted that the analyses contained in this book, brought forth by the fuel crisis, will serve to heighten awareness of important issues. Whilst not all of the 'lessons' will be new, they are strengthened by the insight gained from the events of September 2000.

Glenn Lyons
Kiron Chatterjee

Acknowledgements

The editors wish to thank the Engineering and Physical Sciences Research Council for its sponsorship of the research which prompted the creation of this book. They would also like to acknowledge the Transportation Research Group at the University of Southampton which took the unusual step of underwriting the costs of conducting an emergency survey of the UK fuel crisis in the absence of immediate availability of external funding. The editors' grateful thanks are extended to all the contributors to this volume for their diligence and prompt action to enable the timely conveyance of lessons from the fuel tax protests.

PART I:
AN APPRECIATION OF THE ROOTS OF THE FUEL TAX PROTESTS

1 Fuel Tax Protests and Fuel Rationing – an Historical Perspective

REG HARMAN

Introduction

Shortages are a fact of life: they are also a factor in the economics of human lifestyles. Populations, both human and animal, expand and contract in line with the availability of necessary resources. Sometimes they fight to retain adequate supplies. The territorial ambition that underlies wars often reflects the need to secure sufficient land to underpin the nation's lifestyle: sometimes the nation's leadership promotes such conflicts to keep the citizens happy and therefore keep themselves in power.

By implication shortages in supply become major issues when they affect the key elements of lifestyle. Shortages can reflect a natural shortfall worldwide, a problem in distribution, or a deliberate withholding of supplies from one group by another. Food and shelter have always been basic needs for human beings. In medieval England it was said that even wealthy cities were at any time only six weeks from starvation: the then basic means of distribution and storage meant that even a temporary problem – failure of one local store of grain – could not be overcome because of the time needed to communicate with potential suppliers and move replacements in (if these could be obtained). In modern times such images as television reports from African famines tell us that this remains a threat in some parts of the world, where the poor economy (sometimes exacerbated by regional conflicts) means that food and shelter remain critical elements.

In the modern United Kingdom economy, adequate supplies of food are always available, and there is shelter and clothing for everyone (though the quality and choice available to different groups remain an issue). Other elements have now moved into the category of being essential to current lifestyles, and so have become objects of personal and political debate and dispute. Some examples are extreme: for example, Christmas rarely passes without shortages in some widely fashionable toy occupying newspaper headlines as parents spend time scouring shops for one. Others have entered the national lifestyle in a fundamental way, so that even threatened shortages pose serious problems and bring immediate reactions. The major one of these is petroleum oil, especially for use as motor vehicle fuel.

Petroleum oil has several main uses: as petrol and diesel for cars and other vehicles, fuel for heating and feedstock for many chemical products. The blockades of oil depots in September 2000 brought strong popular support and political reaction because the price of petrol had become such an important issue for a good part of the population. Yet grave concern emerged once people realised that the cuts in supply could harm the economy and society. Both factors reflected the extent to which oil products have become essential, with processes based on it permeating a large part of the national daily life.

This chapter seeks to explain how and why petroleum oil has reached this position over the last century, and to review previous occasions of petrol shortages in the UK. It first considers two complementary trends: the growth in demand and supply of petroleum oil during the twentieth century; and the growth of ownership and use of private motorcars in the UK. Then it goes on to review the four previous occasions when shortages in petroleum fuel have bitten into our national economy and perspective: World War II and its aftermath; the Suez crisis of 1957; the Arab–Israeli war of 1973-74; and the Iran–Iraq war of 1979-80 (the last two are generally described as occasions of 'petroleum shock'). Each case includes reviews of the background, the events and the aftermath. Finally some broad conclusions are drawn.

The review focuses on private cars and petrol availability and prices. Commercial road freight, mostly powered by diesel engines, is also important and has been affected by theses crises, but the principal issue for public and political action has been the effects on private motoring.

Material in the chapter comes from three main sources. Activities on the international stage have been drawn primarily from Yergin (1991), complemented by Jenkins (1989). Statistics come from Institute of

Petroleum (IP) and from Department for Transport, Local Government and the Regions (DTLR) material. Events in the UK during previous shortages have been pieced together from detailed studies (using microfilm sets) into contemporary copies of *The Times* (1957 and 1973-74) and the *Daily Telegraph* (1979-80). Specific references are made to these and other sources where appropriate.

Petroleum Oil and the Twentieth Century Economy

Petroleum oil first seeped out of the ground at least 5000 years ago, and it became a local source of heating and energy. By the Graeco-Roman era it had also become a major weapon ('Greek fire'). However, it remained a largely localised energy source for many centuries.

The late nineteenth century saw the invention of the internal combustion engine, which was developed rapidly as a power source for machinery and road vehicles. The resulting demand for petroleum oil brought rapid creation of a whole system for extraction, refinery and distribution. Supply and demand grew apace, creating effectively a single world market. Because the refining of crude oil into petrol produces a residue of heavy oils, the growth also generated markets for other uses, notably to burn heavy oil in thermal power generation (in place of coal). This aided the economics of developing oil refining and distribution.

Much of the early development took place in the United States of America (USA), which had its own substantial supplies, as did the newly emerging Union of Socialist Soviet Republics (USSR). The nations in North America and the USSR (now the Commonwealth of Independent States, CIS) have continued to supply a major part of the world's oil ever since, but largely for use in their own territories. For most of the world the Middle East states formed effectively the only supply source in the early decades, and they have remained dominant suppliers outside the industrialised countries, with a very influential role in the world market. This has been exploited politically on occasions, especially as the Middle East has remained a world region of considerable political turbulence during the twentieth century. In 1960 the main oil producing states formed the Organisation of Petroleum Exporting Countries (OPEC) to co-ordinate their policies and actions.

Much of the rapid economic expansion in the twentieth century has been supported by efficiencies gained through use of petroleum oil: for internal combustion engines in transport and industry, and by burning oil fuel in power generation and domestic heating. In consequence, the economy has become more and more dependent on oil fuels, especially as they also displaced other energy sources and products. Table 1.1 (Jenkins, 1989; Annual Abstract of Statistics, 2000) sets out consumption for the UK at selected dates during the century. Petrol (motor spirit) was a major part of oil before the 1950s. Since 1971 net national consumption has been much lower, albeit rising slowly; but the amounts processed into petrol, for private motoring, have become more significant again.

Table 1.1 UK consumption of petroleum products 1921-99 (thousand tonnes)

Year	Motor spirit	Middle distillates (incl. diesel fuel)	Fuel oils	Other (incl. aviation fuel)	Total used (excl. refinery fuel)	Motor spirit as per cent of total
1921	1040	814	535	1341	2830	37
1938	4883	1429	1242	1595	8969	54
1951	5454	4423	3418	2822	16117	34
1961	8143	8428	18817	7146	42534	19
1971	14734	20181	38772	16464	90151	16
1981	18718	19184	15656	12698	66256	28
1991	24022	27205	11948	11331	74506	32
1999	21512	32366	2186	14909	70973	30

The overall place of petroleum oil in the UK energy supply is outlined by Table 1.2 (Jenkins, 1989; Annual Abstract of Statistics, 2000). As this shows, oil use reached a peak in total energy use by 1971, natural gas playing an increasingly major role subsequently. What stands out is the decline in the importance of coal, even in recent years.

Three stages can be identified to bringing oil products to the marketplace. The first lies in extracting crude oil from under the ground;

the second in processing it into various products, i.e. refining it, at refineries; the third in distributing the products to the retail outlet. Between these stages the oil has to be transported. The major players in this process have always been two:

- the major international oil companies, who purchase the crude oil, move it to refineries for processing and then distribute oil products; and
- the producer countries, who manage extraction themselves or licence oil companies to do it, in either case levying a charge per barrel.

Table 1.2 UK energy consumption by primary source 1931-99 (per cent)

Year	Petroleum	Coal	Natural gas	Hydro energy	Nuclear energy	Total
1931	5	95	-	-	-	100
1951	11	89	-	-	-	100
1961	27	72	-	1	-	100
1971	45	42	9	1	3	100
1981	35	37	23	1	4	100
1991	36	29	25	1	9	100
1999	32	16	40	2	10	100

The major producers work on the basis of long-term agreements wherever possible, but not all oil is sold on this basis. Both they and other dealers buy and sell oil on the international markets under short term deals, but the resulting 'spot prices' are much more liable to current market fluctuations. Distribution within domestic markets, such as for petrol, is handled by both the major companies and independent distributors, the latter buying from the major companies or through world markets.

Basic world prices (at wellhead) remained broadly stable from 1880 until 1970, even through two World Wars, at around $5 a barrel. The crises of 1973-74 and 1979-80 brought dramatic short-term rises, to over $30 per barrel. Since then there have been significant increases in oil supplied from sources controlled by industrialised countries (e.g. the North Sea). This, together with demand constrained by price levels, has kept the typical price broadly around $20 per barrel; but there have continued to be strong fluctuations over time, especially in 'spot prices'. (All prices are in current cash terms.)

The normal time lag between pumping a barrel of crude oil out of the ground and pumping the petrol made from it out of a petrol pump is about 90 days. In the UK oil companies have generally aimed to maintain stocks for at least 70 days' supply, preferably around 80 days. However, with modern communications where news is rapidly transmitted round the world, problems in source countries can rapidly influence the behaviour of corporations and individuals alike, so that actions by consumers or retailers can change potential problems into real ones, bringing shortages and price rises. This factor was important in the 1970s petrol crises, and has caused shortages of other commodities (the 1979 crisis was preceded by a food shortage in the UK; the 1973 one was preceded by a world-wide shortage of toilet rolls!).

The Motorcar and Travel in the UK

The use of private motorcars (mostly driven by petrol rather than diesel engines) has formed an integral part of the growth of the petroleum economy. It has been underpinned by the expanding supply and at the same time it has driven the need for expanding the supply of fuel. More critically, use of the private car has itself changed the pattern of how and where people live, work and play. A car offers unique freedom to travel any direction and distance (road conditions permitting) at the preferred time. This flexibility moved from a luxury in the early twentieth century to an opportunity during the 1950s and a necessity by the end of the century. Total motoring costs have fallen over time too. Substantial changes in the built environment to accommodate the motorcar have created circumstances that have encouraged and then necessitated further car use. Serious questions on the wisdom of such policies started to be posed from the 1970s, but this had little impact until the late 1990s.

The growth in car ownership and use in the UK over the last century is shown by the key figures for selected years in Table 1.3 (DETR, 2000). In 1951 few people had one; within half a century the number had risen tenfold, to one per household on average (in practice many households do not have one while others have two or three). The average distance driven by each car has risen slowly, but the total level of motoring has gone up hugely and continues to climb.

This growth in motoring has formed a fundamental part of growth in the economy, driven by convenience and by price relative to other goods and services (factors widely explored in literature on transport). Petrol prices reflect both commercial factors and Government taxes; the first petrol tax was imposed in April 1909. However, prices have over the century fluctuated, with long periods of relative stability, interspersed by sharp rises, including the two World Wars (prices fell back after World War I to pre-War levels but continued to rise in the austerity era after World War II). Government also determine the charge for the annual Vehicle Excise Duty (licence). But the availability of petrol and the costs of other elements of motoring (purchase costs, maintenance costs, spare parts) are determined primarily by the oil and motor industries in line with market factors, as for other goods and services.

Table 1.3 Car ownership and use in Great Britain 1901-2001

Year	Private cars (million)	Average cars per household	Vehicle distance (billion km)	Kilometres per vehicle p.a. ('000)
1901	-	n/a	n/a	n/a
1921	0.24	n/a	n/a	n/a
1938	1.94	n/a	27.8	14.3
1951	2.10	0.15	29.3	14.0
1961	5.30	0.33	76.9	14.5
1971	10.44	0.62	165.1	15.8
1981	14.87	0.78	219.5	14.8
1991	19.74	0.97	335.2	17.0
1999	22.79	1.06	380.1	16.7

Garages and filling stations form the main retail elements for users of motor vehicles, but their position is influenced by the actions of the major suppliers. This was particularly evident during the petrol crises discussed in this chapter; they were at the 'front line' of supplying motorists but their ability to meet demand on the forecourt was heavily influenced by decisions of Government and the major oil companies. From the 1960s petrol distribution has followed the general trend in retailing towards larger and more efficient units; smaller filling stations with low turnover have found it increasingly difficult to gain support from the major oil companies or even from the independent suppliers, a tendency accelerated by events of 1973-74 and 1979-80. The number of petrol stations in the UK has fallen substantially over the last thirty years: in 1973 there were about 33,000 (similar to the 1930s level), but this fell to 26,000 by 1979, 20,000 by 1989 and 13,000 by 2000.

World War II

By 1939, when the United Kingdom entered World War II, contingency plans for a possible conflict had been laid, including the need to ration many goods and services. Rationing of petrol for private cars was introduced on 23 September 1939, only 20 days after the declaration of war. In all aspects of life, priority was given to manufacturing and to the armed services, and private motoring was seen as a fringe luxury. Travel was generally constrained, with the widespread motto 'Is your journey really necessary?' (most illustrations being of travel by rail, then the dominant mode for longer journeys). (Material in this section is based primarily on Mercer (1990).)

Rationing covered almost all household goods, and this included basic foodstuffs. Initially difficult, this became much worse in stages as the War entered darker phases. Christmas 1939 was 'not too bad' but by March 1942 real austerity was the order of the day; even basic foodstuffs were in very short supply. Despite some grumbling, the grave nature of the whole situation overshadowed most other views, and people carried on as best they could, from sheer necessity. Although a Black Market developed, rationing was strongly managed and enforced (in April 1944 Ivor Novello, an entertainment 'super-star' of his day, was sentenced to two months' imprisonment for offences connected with petrol rationing).

Not surprisingly, the level of private motoring fell significantly. The ownership of cars also declined; in 1943 the number registered was

718,000, one-third of the 1939 figure, and about the same as 1926. A lot of cars were probably laid up in garages, and available for re-licensing when the situation improved. With the easing of restrictions as hostilities moved towards the end, the number rose again, to 1,487,000 in 1946.

But, the end of War did not spell the end of rationing: the overall impact on the UK economy had been enormous, and priority was given to re-investment. Rationing of many items continued for some years. This era of austerity continued until the Festival of Britain in 1951 marked the return to normality in a growing economy. Petrol rationing ended on 26 May 1950.

1957 – the Suez Crisis

Until 1957 most of the tankers bringing Middle East oil supplies to Western Europe passed through the Suez Canal. This had been built by French concessionaires, and was still run by them in the early 1950s. In April 1956 the emerging nationalist government of Egypt announced that it would not renew the concession. Both France and the UK had retained an armed presence in the designated Suez Canal Zone to ensure safe passage of ships, but the British forces withdrew in June. In July the newly appointed President Nasser nationalised the Canal. In October 1956 Israel launched an attack on Egypt, while the UK and France sent in an invasion force to secure the Canal. In response, a United Nations force was sent in to guard the Canal in mid-November, and to arrange its clearance. (Material in this section is based primarily on Yergin (1991) and contemporary issues of *The Times*.)

This crisis precipitated a sudden shortage in petroleum oil, largely reflecting the much longer time for oil to be transported round the Cape. In consequence, prices for crude oil rose rapidly, with strong fluctuations for 'spot prices'. In the UK these factors reduced supplies in the short term by about 25 per cent.

In response, the UK Government announced on 29 November 1956 that rationing of motor spirit (petrol) and DERV (diesel engine fuel) would come into force on 17 December 1956. The scheme set monthly quotas for types of vehicle. For cars and motor cycles, this varied according to engine size (then measured in both HP or cubic capacity (cc)), but was set to allow travel of 200 miles per month (typical car use then was stated as

700 miles per month). For goods vehicles the allowance was 2 gallons of petrol per half ton unladen weight, or DERV to enable the same mileage. Supplementary allowances could be claimed for essential purposes. Operators of road public passenger transport had their supplies reduced by 10 per cent. The declared aim was to maintain supplies for industry (Institute of Petroleum, 1957).

By January 1957 supplementary allowances, allowing 300 miles per month in total, had been agreed for about two million cars and motorcycles. These covered self-employed people, and employees who could only reach work by car / motorcycle, whose difficulties would therefore cause loss of production. They were administered by regional rationing offices set up by Government. This did not satisfy the United Commercial Travellers Association, who in their campaign for increases pointed out that commercial travellers averaged 1000 miles a month. Many cases of distinct hardship were reported, mostly in rural areas, where supplementary coupons were provided by the volunteer Women's Royal Voluntary Service (WRVS). There was general agreement that most use of cars was for business and employment; 'pleasure motoring' was estimated to account for one-fifth of all car mileage. Fears were expressed over the problems which the reduced supplies were causing road haulage; Government agreed an increased allowance for goods vehicles.

By March 1957 stocks were building up again. On 17 April the basic ration for cars was increased to 300 miles per month (with supplementary allowances on top). With stocks continuing to increase, rationing ended on 15 May 1957. Meanwhile British Railways reported increases in passengers, as did London Transport and other bus operators.

Government also increased the motor fuel tax on 5 December 1956, from 2s 6d (the level since January 1952) to 3s 6d, to restrain consumption. It was returned 2s 6d on 10 April 1957. Most bus companies increased fares to recoup the higher fuel cost; some did not reduce fares when the tax level fell, despite criticism from companies that did reduce fares again.

There were no serious repercussions for motoring in the UK. Developments in markets and technology were already leading towards smaller and more economical cars, such as the innovative BMC Mini (introduced 1959) and succeeding 'super-minis'. Car ownership, 3707,500 in 1957, continued to rise. Car mileage in 1957 was 2.2 per cent lower than in 1956, but it then rose by 22.6 per cent in 1958, and continued to rise (DETR, 2000). The whole economy also increased steadily, with greater dependence on oil (an issue developed in the next section). However, the Government did gain support for an increased programme to build nuclear power stations. The major oil suppliers hastened to bring into service very large ocean going tankers ('supertankers') that could carry large quantities of crude oil economically over long distances, to bring supplies for the UK, and the rest of Western Europe, round the Cape.

1973-74 – the Arab-Israeli Conflict

The period from the late 1950s to the early 1970s was one of rapid economic growth in the UK, as for most of Europe, with new materials and technology playing an important role. This era was summed up by two contemporary prime ministers as 'you've never had it so good' (Macmillan) and 'the white-hot heat of technology' (Wilson). 11,738,000 cars were registered in 1973, over three times as many as at the 1957 Suez crisis. Travel by bus and coach, and by rail, was in decline. Oil played a large part throughout the economy, and growth in its use was accompanied by decline in coal (Table 1.2), a trend broadly reflected by UK policies on energy and the economy. This caused a growing dispute between coal miners represented by the National Union of Mineworkers and the publicly owned National Coal Board, which in 1973 led to an all-out strike, with real conflict at mine gates. (Material in this section is based primarily on Yergin (1991) and contemporary issues of *The Times*.)

Conflict in the Middle East then brought a further sharp disruption. In 1967 Israel had launched an unexpected attack on Egypt (the Six-day War). In early October 1973 Egypt launched a similar attack on Israel, during the Jewish holy days of Yom Kippur. The Yom Kippur War lasted three weeks and left a state of serious unrest in the Middle East. The leading oil producers, through OPEC, decided to reduce their production, causing an immediate pressure on oil supplies around the world.

Uncertainty over the Middle East situation had already led Government to consider rationing; coupons and plans to distribute them were prepared in August 1973. In October the Government set up an emergency Committee, chaired by the Leader of the House of Commons, to propose advisory measures on fuel reduction and to oversee rationing if needed. Books of ration coupons were sent out to all post offices and sub post offices in the first week of November and distributed to the public between 29 November and 12 December. These gave a basic allowance to all car owners (regardless of car engine size); but a higher allowance was given to people in the medical, welfare and religious professions, and to registered disabled people in employment. Others could claim a supplementary allowance by application to regional centres. Coupons for commercial goods vehicle operators (HGV O licence holders) were issued by Crown post offices, and for bus and coach (PSV) operators by the Traffic Commissioners.

By early November total UK oil stocks still stood at 79 days' supply (about the normal level) but they were starting to fall sharply. Esso started to ration supplies to its petrol stations, Shell and BP reduced the octane ratings of their petrol to obtain more petrol per barrel of crude oil. They all cut supplies to the independent distributors, who were therefore forced to buy more at very high 'spot prices'. The effects soon showed on petrol station forecourts. Some ran out and closed until new supplies arrived. Others introduced limits, such as 4 gallons per motorist, or served only credit customers (largely business users) or regular customers who had registered; panic buying started. Shell and BP encouraged drivers to observe lower speeds, but they also challenged Government for not giving a stronger lead. Government in turn complained that the major distributors were not allocating fuel fairly in some cases. By 23 November total stocks had fallen to about 65 days (below the accepted minimum of 70 days). The main independent suppliers, Blue Star and Heron, now had 20 per cent less supplies than normal, and were encountering serious problems, disturbances on their forecourts matched with falling profit margins; they called for rationing to be introduced immediately. Hoarding of petrol at home in 5- and 6-gallon cans was now becoming common, and petrol theft was being reported, and occasionally prosecuted: motorists hurried en masse to buy lockable filler caps for their cars' petrol tanks.

By early December the petrol shortage was reaching crisis point. On Sunday 2 December only one in ten petrol stations opened, and most of them closed by lunchtime. By mid-week, panic buying had appeared in London and the South East, with long queues at open garages, including one a mile long in Borough High Street. Disturbances occurred over priorities and perceived queue-jumping; the police had to attend a number of incidents, and occasionally detained motorists to keep the peace. (One Opposition party Whip, refused petrol because he was not a credit customer, blocked the entrance to the garage in question, shouting 'pump power to the people!' to queuing motorists.) The shortages were accentuated by motorists visiting petrol stations more regularly in order to keep their tanks topped up, and sometimes driving long distances to refill. Some garages now limited purchases to 50p per visit, worth less than 2 gallons: the AA pointed out that this forced motorists to drive further and hence waste fuel. Two letters in *The Times* (7 December) calculated that, by keeping their tanks topped up, motorists were transferring stocks from petrol stations, hence worsening the shortages and adding to the panic.

Meanwhile, at midnight on 13 November 1973, the Government declared a State of Emergency, allowing it to promulgate directives and regulations. On 20 November Government directed the oil companies to cut supplies of all petroleum products, including petrol and diesel, by 10 per cent, and forbad the purchase of oil (storage) cans. On 4 December Government ordered the oil companies to divert more oil to power stations for electricity generation (coal was now severely restricted by the miners' strike). On 7 December it imposed a speed limit of 50 mph on the whole road network (having first recommended this on 20 November).

But exhortation remained the Government's principal means of dealing with the crisis. Government regularly pointed out that the total cuts in petrol supplies were relatively small, so scope existed for motorists to make savings, especially in private non-essential motoring. It encouraged motorists not to drive on Sundays and to share cars wherever possible. On 27 November, as the crisis deepened, Government urged petrol stations to give priority to people in those categories who were due additional allowances in the petrol ration coupons yet to be distributed!

These exhortations were backed up by the motoring organisations, the Automobile Association (AA) and the Royal Automobile Club (RAC). On 21 November they issued detailed advice, pointing out that driving at 50 mph instead of 70 mph would gain an extra 10 miles per gallon and offered guidance on better driving, better maintenance, and the need to switch off the engine if stopped in a jam. They both reported that motorists were generally following the advice given, with reduced speeds, and cuts in Sunday driving; except that traffic levels in the South East did not diminish. The police confirmed that speeds were generally reduced, although some drivers continued to drive at very high speed on the M1. But the AA and RAC also stated on 27 November that Government should make the key decisions, as Government were best placed to identify available stocks and future prospects.

Correspondence in *The Times* looked at the needs of particular groups, such as the disabled, the impact of rationing on the majority of car users, and wider issues. A few correspondents pointed out that public transport needed a major Government-led development programme to be effective, and that it had been undermined by failures to redevelop towns and cities to support it (Milton Keynes was cited as a prime example). On 1 December the Chairman of C & J Clark Ltd of Street (the major Somerset shoe manufacturer) pointed out that major changes in the economies and lifestyles of the many South West market towns had taken place, so that:

> a whole network of private car transport has taken the place of public transport, both road and rail, and, for better or worse, it is now an integral factor of the rural economy … The car has become an essential factor of the rural economy network.

On 8 December 1973 *The Times* reported a survey carried out for it on the handling of the fuel crisis. Of the respondents:

- 70 per cent agreed with the Government's view that the shortages were serious, 23 per cent did not;
- 51 per cent thought the Government had handled it well, 30 per cent badly;
- 53 per cent thought that most people would follow Government's advice on reducing use without the force of law but 40 per cent thought that they would not; and
- 90 per cent agreed that rationing through use of coupons was the fairest way to ensure all motorists got a supply of petrol, 7 per cent disagreed.

Yet within a very few days the crisis on the forecourts was over. The Secretary General of OPEC had declared on 29 November that producing countries would maintain supplies for the UK. By 7 December stock shortages were easing, and more petrol was available; albeit sometimes at a high price. Petrol supplies gradually increased throughout December, though travel over the Christmas holiday period was lower than expected. By this time however motorists, and everyone else, had graver matters on their minds. On 14 December the Prime Minster announced that from 31 December 1973 all industrial and other production would be restricted to three days per week, in order to stabilise the national economy, which had been rocked by the combined effects of the oil, coal and electricity shortages, and by strikes in other major industries. Share prices were tumbling; industry was braced for large job losses. The 'winter of discontent' was casting its cold spell over UK society (it would prove a factor in the change of Government at the election of May 1974).

In effect the queues at petrol stations formed part of a wider shock throughout the economy that was to start a review of UK energy policy. This led to a reduction in total dependence on oil (encouraged by the higher international prices) and transfer to other fuels, especially the supplies of natural gas from the North Sea. The North Sea fields were proving rich in petroleum oil too: one of several new oilfields brought on stream by the keenness of the industrialised world to discover new sources away from the Middle East. A long-term vision was also being promulgated by the emerging environmental study and pressure groups; but their calls to establish economic sustainability would have little immediate impact.

While supplies of petrol returned to something like normal during 1974, prices did not. In October 1973 a gallon of 4 star cost 37.5p (including 22.5p VAT). In February 1974 it cost (with the same VAT) 50p. By November 1974 (with 22.5p VAT and a 25 per cent petrol tax) it cost 62.5p, but the following month (with the same tax charges) it was 72.5p. By November 1976 (with VAT at 30p and a 12.5 per cent petrol tax) it was 82p. In other words, the pump price more than doubled in just over three years, but with most of the rise in 1974 (Institute of Petroleum statistics). Yet car ownership and use continued to increase. Total miles driven in 1974 were 2.2 per cent lower than 1973, but they rose by 9 per cent in 1975 and growth increased from 1976 to 1978 (DETR, 2000).

1979-80 – The Iran-Iraq Conflict

1979 was the 'summer of shortages'. The year started with a mini-crisis in petrol supplies, caused through a strike by tanker drivers. This was part of a rising dispute across the road haulage industry, which blocked ports and led to serious shortages in supplies of materials to industry and food to shops (probably factors in the change of Government at the election of May 1979). By the second week of January the strike had brought a serious shortage of petrol at petrol stations, to the extent that by the middle of the week many closed down. The London area saw few problems, but the Midlands, North West and Glasgow areas were seriously affected: Scratchwood services on the M1 ran out completely. Those which remained open limited drivers to £3 of petrol, equal to 3.75 gallons of 4 star; though some were charging £5 for 3 gallons. Massive queues built up, disturbances broke out, and Government considered calling in the Army to distribute petrol. However the strike was soon settled, drivers finally returning to work on 15 January. By now people were queuing for food in the shops (and bemoaning a reduction in choice) and business leaders were starting to fear serious loss of jobs. (Material in this section is based primarily on Yergin (1991) and contemporary issues of the *Daily Telegraph* (*The Times* itself was closed by disputes for most of this period).)

Meanwhile, further problems were building up in the Middle East. In late 1978 the Iranian government, under the leadership of the Shah, had suspended exports of crude oil, because of its own internal disputes, which led to revolution in February 1979. Although Saudi Arabia temporarily raised production by a quarter, total Middle East output was significantly reduced. This brought a scramble by the major international oil suppliers to secure crude oil, during which existing structures and agreements were abandoned. BP was heavily dependent on Iranian supplies; to ensure sufficient supplies, it sought out other sources, reduced supplies to all customers (in the UK and elsewhere), and refused to renew contracts. Other companies began to follow: 'and thus the dominoes began to fall' (Yergin, 1991). The governments of industrialised nations failed to take cohesive action and world-wide panic ensued. The main oil exporting countries, co-ordinated through OPEC, realised their opportunity, and in March 1979 reduced consumption and increased the standard price substantially.

Shortages at petrol stations followed. In April the main oil companies advised that they would limit deliveries to 1978 levels; the independent suppliers were facing severe problems. By the middle of May most filling stations except those of BP were receiving between 5 per cent and 20 per cent less petrol, and prices at the pump were varying widely: between 78p and 98p a gallon, according to a survey by the RAC, with prices up to

£1.10 in some rural and independent filling stations. By the end of May long queues were building up on occasions at filling stations, and police were being called out from time to time to deal with disturbances. People were started to bring in 5-gallon cans to fill up for storage, but this was stopped by the police. Concerns were voiced about the impact over the Spring Bank Holiday, but poor weather and general uncertainty discouraged motoring, and few problems were encountered, despite the fact that most filling stations closed.

Serious problems continued throughout June and much of July, with the petrol allocations remaining uncertain. BP reduced supplies by 5 per cent, and Esso and Shell made further cuts; all the oil companies were concerned about maintaining stocks of oil for winter. Most filling stations closed for Sundays. By mid-June some were imposing limits of £1 (about a gallon), and prices were continuing to rise, up to £1.30 a gallon maximum. To obtain petrol the independent retailers had to pay very high 'spot prices', which were reflected in their pump prices. These high prices were then matched by other retailers; all petrol stations were suffering with low sales and poor profit margins, posing problems for their survival. Some motorway filling stations had shortages, leading to queues. But the higher prices and general air of uncertainty did reduce demand to some extent. Petrol thefts were also up significantly, and again there was a rush to buy lockable petrol tank filler caps.

By early July the major oil companies were securing more oil on the international markets (albeit at a high price), and on 6 July they advised the Government that there were not likely to be any problems with winter supplies. Petrol was being distributed more easily and panic buying ceased. On 23 July the AA and RAC reported that motoring was becoming easier; on 15 August they reported that tank-topping was over, and prices were stabilising at around £1.20 per gallon. No problems were reported at the August Bank Holiday. Meanwhile the motor industry reported on 9 July that car sales for the year were at a record level!

Particularly serious shortages were felt along the Sussex Coast, which experienced panic buying and long queues from late May. Subsequently the West Country and East Anglia also encountered serious shortages, partly reflecting the large number of small petrol stations linked to independent retail groups in these more rural regions. The AA and RAC expressed particular concern about the impact on holiday and leisure motoring. The English and Scottish Tourist Boards reported on 5 July that demand for accommodation in British holiday resorts was down about 5 per cent, reflecting worries about petrol shortages: they complained that the media and motoring organisations had overstated shortages in holiday areas.

The newly elected Government had started to review the situation from May as it developed, and considered the need for rationing, but ruled this out by late May. Government maintained throughout the crisis that rationing was unnecessary unless further major problems arose in foreign (oil producing) countries. The oil companies criticised Government for failing to act more positively to secure national oil supplies. There was widespread criticism that 45 per cent of the production of the British National Oil Corporation (BNOC), the company responsible for exploiting North Sea production, was exported: Government pointed out that BNOC had international contracts to meet.

Government continued to urge careful driving, pointing out that savings of 5 per cent cut in oil use, equal to 7 per cent cut in demand for oil products such as petrol, would be sufficient. The AA and RAC issued guidance on efficient driving. The Government's target was largely achieved during the crisis period, but through the rapid price increases and shortages at the pumps. Total miles driven in 1979 were 0.5 per cent lower than for 1978.

Government liased with the oil companies throughout the crisis period, and particularly over priorities for distributing petrol and other oil products; but the role of the Department of Energy was described as acting 'only in an information-gathering and advice-giving capacity' (*Daily Telegraph* 22 June 1979). In practice the responsibility for action fell mostly on the oil companies, who allocated supplies to their filling stations on a daily basis. In doing so, they had to take into account the cost of supply and potential return. Large garages in urban areas and on main trunk routes were easier to supply, sold much more petrol (and diesel fuel) and could survive on lower retail margins per gallon. More remote rural petrol stations were expensive to supply and sold relatively low amounts. Some reports suggested that the major oil companies were giving priority to larger petrol stations. They continued the process of rationalising their networks, cutting out smaller more remote petrol stations.

After the 1979 crisis sharp rises in oil prices occurred, complicated by further Middle East disputes. In July 1979 the Shah left Iran, and Ayatollah Khomeni returned to lead a revolutionary government. By

November this government was in major dispute with the USA, leading to sanctions against it from both the USA and the European Union in April 1980. In September 1980 Iraq attacked Iran, seriously damaging the major Abadan refinery; in October Syria sided with Iran. By December 1980 the OPEC guide price was $32 a barrel, $8 higher than a year earlier. However, in January 1981 the USA deregulated its own crude oil prices, leading to widespread price reductions across the world.

For the UK motorist these events brought two years of substantial rises in the typical price of petrol. In April 1979 4 star petrol was 88p a gallon; by March 1980 it was £1.37 a gallon, by March 1981 £1.57 a gallon. This partly reflected a doubling of duty, from 30p per gallon to 62.83p per gallon, with VAT also rising from 12.5 per cent to 15 per cent. Duty continued to rise during the 1980s, reaching 92.92p per gallon by April 1988, but the typical price for 4 star petrol was then £1.68; though it had been as high as £2.05 in April 1985 (Institute of Petroleum statistics).

None of this stopped growth in car ownership and use. Total mileage in 1981 was up 9.0 per cent higher than 1979; it continued to climb steadily through the 1980s and also through the 1990s; and by 1999 it was 73 per cent higher than 1981 (DETR, 2000). This against the background of general stability in long term oil prices. In March 1983 failure by OPEC members to agree production and pricing policy had led to the standard price falling to $29 a barrel. Since then changing policies by industrialised nations, especially further development of new oil fields, have kept prices more generally stable. There have been some interruptions, notably the 1990 invasion of Kuwait by Iraq; events such as these have pushed up 'spot prices' sharply on occasions, bringing some variations in UK petrol prices.

Conclusions

The motor car has become a fundamental item of UK society very quickly: the main part of the growth in ownership and use covered barely a generation. It relies on supplies of petroleum oil, a material which has also come to dominate the whole UK economy over the last century. Although the main actors – the producing countries and the large oil refining and distributing companies – are all major organisations, the constant pressure to maintain supplies at existing levels means that any perturbation in the overall chain of supply can cause painful ripples throughout. The protests of September 2000 showed just how rapid and serious the effects could be.

The twentieth century has seen four previous occasions when supplies of fuel for motoring in the UK were interrupted. Successive events saw a changing pattern in both Government actions and in the responses of the motoring public, reflecting the stages in the importance of the car. In World War II, Government imposed severe rationing, leading to widespread reduction in car use, as part of much wider rationing, during a long period of grave emergency. The 1957 Suez crisis followed only a few years after the end of wartime rationing; again, Government imposed severe rationing from the start and kept it for six months, until fuel supplies were fully restored. In both cases some concerns were expressed by business users, but car ownership remained low.

However, by 1973 car ownership had become much more significant: when Middle East conflict threatened oil supplies, the Government prepared for possible rationing but never imposed it, preferring to act through exhortation and some regulation to reduce use of petrol (and other oil products). When a repeat run occurred in 1979, Government did not even prepare for rationing or introduce regulations, relying entirely on exhortation and guidance. No attempt was made to introduce rationing in 2000; although the speed and cause of the fuel shortages then did not leave much scope for this. It appears that, as car ownership and use moved from minority element to a basic element in lifestyles, Government has become increasingly reluctant to act in any way that might seem to impose restraint. Yet in the 1970s cases both oil suppliers and motoring organisations urged them to do so, and a survey of motorists confirmed support for such action. Furthermore, Government was apparently reluctant to challenge the major international oil companies.

In the absence of firm Government leadership, major suppliers effectively controlled distribution of petrol (and other oil products), through allocations to their retail outlets and through sales in other markets. But these companies were not responsible for all petrol stations, and had to maximise their own position in the crisis: so petrol stations supplied by the independent retail groups suffered badly, through shortages or through charging much higher prices to meet the very high costs of alternative sources of supply. Moreover, even within the main supplier groups many petrol stations set their own constraints and charges during the crisis period, in response to immediate market conditions.

In practice the total reduction in petrol supplies during the 1970s 'petroleum shocks' was relatively small, as Government kept pointing out. The problems were caused primarily by the behaviour of motorists seeking to keep their tanks topped up at all times for fear that they might not find any petrol the next day; a factor also experienced in the September 2000 crisis. Given the continuing uncertainties during the short periods in question, this was a logical action for those who relied on their car for their everyday activities. Rationing would have largely eliminated these problems, by giving people some certainty within which to plan their activities and priorities. Sunday motoring was reduced in 1973 and 1979, suggesting that people did give some thought to priorities and cut back on the 'pleasure motoring' that had been criticised in 1957. However, the reduction in leisure motoring hit the tourist and leisure industry, to the concern of those involved with it.

Yet the 1970s 'petroleum shocks' made barely a dent in the growth of motoring in the UK. Even in the short term people carried on using their cars, despite the very substantial increases in prices, especially in 1979. This indicates very low price elasticity: access to jobs and activities by car had become fundamental to the majority of people's lifestyles. The 'petroleum shocks' caused a dent in the UK economy, which was very heavily dependent on oil at the start of the decade. By the early 1980s the pattern of energy supply was changing; but in total it remains even more heavily dependent on the combination of petroleum oil and gas.

Since 1980 the supply of petroleum oil has been relatively stable, as has the supply at the pumps for an ever increasing fleet of cars. There is no guarantee that this will continue for ever; indeed, general economic and political history suggests that sooner or later continuing growth in demand will run up against an ultimate limit of supply. The experiences of winter 1973, summer 1979 and September 2000 demonstrate how a car (and oil) dependent society will react; though they also throw up some useful pointers for the future.

First, when Government did take action, especially through rationing, this was respected. Admittedly, World War II and the Suez crisis were in an era of low car ownership and different culture. But *The Times* survey of 8 December 1973 showed that most motorists expected Government to take critical decisions, including the imposition of rationing, as they felt Government was best placed to take an overview of the situation and act in the country's best interests. Social and cultural attitudes have probably moved on since 1973, but sound management of crises by Government on a clear and reasonably equitable basis is likely to remain acceptable to most people. Voluntary codes are not accepted by everyone, as the 1973 and 1979 crises showed; indeed, both brief critical periods of unrest and uncertainty were caused primarily by the behaviour of motorists. The commercial bodies involved, such as petrol producers and distributors, were heavily influenced by their position in both international oil markets and national fuel distribution markets, and had to put their shareholders' interests first to ensure survival.

Second, any management of such crises must be framed within the wider consequences of any restraint in car use. The more that cars are an important part of everyday social and economic behaviour, the more that policies to limit their use must address all implications. It is not possible to solve all problems. But understanding must be shown, and issues dealt with in all complementary policy areas. In 1957, 1973, 1979 and 2000 little or no attempt was made to manage travel demand locally or to prioritise public transport.

So, third and finally, Government need to address fuel issues as part of the overall transport picture, on the principle of 'joined-up thinking'. There are other reasons for considering constraints on increased car use, such as better quality towns and cities, improved access for everyone to goods and services, achieving a cleaner environment, and better health and security. To achieve these wide goals, it is necessary to be well informed and to look well ahead. Transport and related polices ought to address the issue of future fuel shortages, perhaps through developing scenarios to test the robustness of future plans against them. The 1970s crises showed that car use may be temporarily reduced through a sharp downturn in the economy: but that does not address the central goal of improving people's lives. The car is a major factor in transport and lifestyles, albeit not the only one. Transport policy formulation must recognise this factor – but not be dominated by it.

References

Annual Abstract of Statistics, 2000. TSO, London.

Daily Telegraph, London.

Department of the Environment, Transport and the Regions (2000). *Transport Statistics Great Britain: 2000 Edition*. TSO, London.

Institute of Petroleum (1957). *Institute of Petroleum Review, Vol.11*.

Jenkins, G. (Ed) (1989). *Oil Economist's Handbook (5th edition)*. Elsevier Applied Science Publishers, London.

Mercer, D. (Ed) (1990). *Chronicles of the Second World War*. Longman, London.

The Times, London.

Yergin, D. (1991). *The Prize (The epic quest for oil, money and power)*. Simon and Schuster, London.

2 Pro-Car or Anti-Car? 'Environment', 'Economy' and 'Liberty' in UK Transport Debates

PAUL ROSEN

Introduction

The fuel tax protests of autumn 2000 strikingly expressed tensions and uncertainties in attitudes in the UK that had been emerging for some time at the intersection of transport, the economy and the environment. Since the late 1980s, transport policy has shifted from a focus on building roads in order to meet demand, to a 'demand management' approach that seeks to limit the need for new road capacity by managing the use of roads more effectively (Owens, 1995). This more recent approach is based on arguments that question both the capacity of our roads to cope with projected traffic growth (SACTRA, 1994) and the environmental sustainability of increased road traffic (Royal Commission on Environmental Pollution, 1995). The latter part of this argument has been underlined by a succession of high profile environmental protests around transport, mostly in the form of occupations of new road development sites (Wall, 1999; Brass and Koziell, 1997).

More recently, though, a new perspective has also become prominent which places other issues – notably the economic costs and personal liberty associated with the car – above environmental concerns. For advocates of this perspective, their priorities are the rights and needs of motorists, and they often regard environmentalists, transport professionals and politicians who argue for reduced car usage as 'anti-car' (Dunn, 1998; Williams, 2000). This lobby, which perhaps warrants a parallel label of 'pro-car', shares many ideas and personnel with other groups such as the Countryside Alliance (which campaigns against Government proposals to outlaw blood sports) that have been formed in opposition to the policies of the New

Labour government elected in 1997. By far the most visible manifestation of this opposition have been the fuel blockades that sparked the September 2000 crisis, initiated by farmers and road hauliers.

This chapter traces the interplay of these so-called pro- and anti-car perspectives in British transport policy over the last decade. The following section examines the policy and cultural changes around transport and the environment that preceded the events of 2000, focusing especially on anti-roads activism in the 1990s. The next two sections concentrate on the perspectives of the pro-car advocates who preceded – and some of whom initiated – the 2000 fuel tax protests. These sections examine the extent to which discourses around the environment, economic costs, and the liberty of drivers were drawn on by politicians, the press and the protesters both prior to and during the fuel crisis itself. The chapter concludes by considering what these themes tell us about the competing agendas of the various parties to the dispute. It is argued that discourses around the environment, economic costs and the freedom of the individual are of great importance in shaping how transport issues are contested by different interests. Recognising this point will be crucial to both policymakers and protesters in any future transport dispute.

Transport and Sustainability: Environmental Activism and Environmental Policy

Concerns about the environmental impacts of transport can be traced back to the emergence of the current environmental movement in the 1960s and 1970s. Key early texts from within this movement highlighted the threats to the natural and social environments caused by modern technology, including the car (Dunn, 1998). The same period saw the formation of high profile environmental campaigning organisations such as Friends of the Earth and Greenpeace, as well as the forerunner of the British Green Party. Such developments can be set against a wider context of the predominantly youth-oriented 'counterculture' that questioned and challenged mainstream society in myriad ways (Roszack, 1971).

Dunn (1998) argues that this counterculture always included a component opposed to the car. Dealing with a predominantly American context, he traces the emergence of what he calls a 'policy vanguard' of the 'enemies of the automobile'. This includes writers, activists and policymakers who have sought to associate 'mass automobility' with threats to the global environment, with planning trends that have prioritised suburbanisation and the rise of shopping malls at the expense of community life, and with car-dependence being subsidised at the expense of public transportation. In a British context, too, it is possible to trace 'anti-car' activism in the 1990s back – in part at least – to this earlier period (McKay, 1996; Wall, 1999).

However, whilst concerns about transport have remained a consistent feature of environmental discourse, it was only in the late 1980s that they became a defining component of it. A key turning point here was the publication in 1987 of the Brundtland Report (World Commission on Environment and Development, 1987), which popularised the concept of 'sustainable development'. Transport has been a significant strand in the sustainability agenda that has followed since – evident, for example, in *Agenda 21*, the international sustainability document adopted at the 1992 Rio Summit (Johnson, 1993). This addresses resource depletion and environmental damage caused by both the production and consumption of motor vehicles, as well as the patterns of urban design that encourage travelling, and the impact of transport on urban populations. It calls for action on these issues in both the developed and developing worlds via a combination of technological, policy and fiscal measures.

Publications such as the Brundtland Report and *Agenda 21* set the framework for a shift in British government policy towards the environment in general, and specifically in relation to transport. The 1990 White Paper on sustainable development, *This Common Inheritance* (Department of the Environment, 1990)[1] framed the issue of transport in terms primarily of market forces responding to demand for environmentally less damaging forms of transport, underpinned by a small degree of regulation to encourage modal shift away from the private car. This was overshadowed at the time, however, by the Government's massive commitment to roadbuilding – set out in its policy statement, *Roads to Prosperity* (Department of Transport, 1989) – and Prime Minister Margaret Thatcher's championing of 'the great car economy' (Media Natura, 1990).

For environmentalists, this position was expected to lead directly to environmental and social degradation, in addition to the impacts of traffic growth on health and local communities that were seen to result from streets becoming less accessible to non-motorised users (Elkin et al., 1991). This outlook is exemplified in the response of the Green Party to the Government's roads policy, which it renamed 'the road to insanity' (Green Party, 1991). The Greens' own policy statement summed up 'the transport problem' thus:

> Transport is in a mess – a crisis. Traffic congestion is wrecking social and street life in many of our towns. Noise pollution and air pollution are making roads and cities unpleasant. Diseases related to car pollution, such as asthma, are on the increase. Residential areas are increasingly becoming commuter car parks. (ibid.: 1)

This kind of perspective reflects the way transport at this time was beginning to take a more prominent position within environmental campaigning, and most notably within the growing world of environmental 'direct action'. The latter, which became increasingly significant during the mid-1990s, coincided – and to some extent helped influence – a gradual shift in policy towards giving greater priority to the environmental impacts of transport. In contrast with the sentiments of the 1989 road programme, the mid-1990s were marked by the publication of several notable documents. The 1994 report of the Standing Advisory Committee on Trunk Road Assessment (SACTRA, 1994) highlighted the traffic generated by new trunk roads; the Royal Commission on Environment and Pollution (1995) raised concerns about the negative environmental impacts of transport; new planning guidance aimed to encourage access for non-car modes to new developments and to prioritise town centre over out-of-town shopping (Departments of Environment and Transport, 1994; Department of the Environment, 1996); and against the background of all these, the then Transport Minister, Dr Brian Mawhinney, set in motion a 'Great Transport Debate' which resulted in the publication of a Government policy statement in 1996 (Department of Transport, 1996) that partially endorsed arguments for more sustainable transport and a modal shift away from the private car. The new Labour government's transport White Paper of 1998 (DETR, 1998) consequently brought these issues more firmly within Government policy, focused especially around the notion of 'integrated transport'. Most notably, the White Paper built to a significant extent on thinking which had already been developing under the Conservatives.

This transformation of policy was set, as indicated above, against a background of increasingly militant environmental activism, whose origins and progress have been traced in detail by Wall (1999). A key stimulus to this activism was the growth during the late 1980s of public concern about environmental issues, expressed for example in the Green Party's unprecedented 15 per cent share of the vote in the 1989 European elections. This was followed, however, by a struggle among British Greens between the Party's pragmatist ('realist') and radical ('fundamentalist') wings. The temporary dominance of the former after 1989 left many people looking for new ways of expressing their concerns (ibid.).

One such expression drew inspiration from the American Earth First! (EF!) movement, focused around preventing development in 'wilderness' areas. A small network of EF! groups was established in Britain starting in 1991, initially concerned with opposing the destruction of developing world rainforest areas. However, opposition to the government's road building programme quickly provided a new focus that was much closer to home – the destruction of green spaces, wildlife habitats and homes in order to build new roads (ibid.). In subsequent years, EF! groups teamed up with a variety of other campaigners to oppose particular roadbuilding projects. These included local campaigners concerned about their local environment who often invited EF! to help them establish a protest, as well as those from different segments of the Green movement who shared many of the aims and strategies of EF! These combined campaigns and campaigners staged a series of anti-roads protests, especially during the period from around 1993-7, moving from one site to another as the Government's roadbuilding programme progressed. The initial catalyst was the planned M3 extension at Twyford Down, which spurred a great deal of the mythology – particularly around the 'Dongas tribe' – that went on to sustain the identity of anti-roads protesters. The last of this main phase of protests took place at the A30 in Devon. As Wall writes, the evictions here:

> marked the end of a long phase of protest. The programme of road construction had been scaled down and new targets attracted green direct action attention. (1999: 92)

The anti-roads protesters adopted a variety of what Wall terms 'repertoires of action', drawing on lessons learned from members' previous experiences in other campaigns and movements. In particular, their activities encompassed a range of protest strategies under the overall banner of 'non-violent direct action' (NVDA). Much of this was learned from the peace, women's and animal rights movements, and included obstructing access to development sites, passive resistance in the face of forced removals by security guards, and – most effectively of all – putting oneself in dangerous situations such as tunnels or treetop walkways; removing protesters from these settings could only be carried out by trained personnel. In addition to this, there was some targeted destruction of property belonging to construction firms.

This range of 'repertoires' gained the protesters considerable media attention, and coalesced into a style of action that contrasted sharply with the more constitutional mainstream environmental organisations established in the 1970s (Rootes, 1997). Within EF!, decision making was consensual rather than centralised, and the movement was structured around networks of very localised groups rather than the hierarchical structures of Greenpeace or Friends of the Earth. This allowed for very fast communication between activists, and also made it difficult to identify any ultimate 'leader' behind actions. There was a prominent use of the new communications technologies appearing at this time – most notably mobile telephones, computers and the internet (McKay, 1996). This facilitated a fast mobilisation of support, and also the wide distribution of publicity, whenever an action took place (ibid.; Wall, 1999).

It is important to note that the apparent homogeneity of environmental protesters – especially as presented in the media – masked a great variety of perspectives, reflecting in part the disparate origins of participants. NVDA itself has been a cause of great debate within direct action circles (Hart, 1997), and a common distinction was made within EF! between those preferring 'fluffy' (i.e. peaceful) as opposed to 'spikey' (i.e. conflictual) activities. That said, the effectiveness of these protests in highlighting the issues lies to a great extent in their ability to bring together contrasting perspectives; the tensions between them were temporarily rendered secondary to the objectives in hand. Nevertheless, there was also much scope for disagreement among participants over tactics or even overall objectives, especially between members of groups such as EF! and more traditional organisations taking part in the same action.

In fact, EF! was not the only newer group involved in such actions. During this same period, other groups also were organising transport-related protests, using similarly high profile tactics. Reclaim The Streets (RTS) had first been established during the 1970s as a means of highlighting the noise and pollution caused by city traffic (ibid.), and this was revived during the 1990s to great effect in a wide range of city contexts. RTS actions block off small sections of city roadways for a street party, thus preventing traffic access, and again use a panoply of direct action techniques learned from past activities. Critical Mass bicycle rides have similar objectives, but block traffic by bringing together large numbers of slow moving cyclists. Both RTS and Critical Mass have become international tools of protest against the dominance of car traffic, and have also formed part of a somewhat refocused repertoire of protest strategies since the late 1990s concerned with challenging global capitalism[2]. This shifting focus from transport to broader concerns highlights the flexibility of contemporary social movements, and the lack of structure compared to earlier movements (Melucci, 1989). Membership of groups such as EF!, RTS and Critical Mass is fluid and multiple – an individual might be a member simultaneously of several groups, whilst the banner under which a particular action takes place will not necessarily wholly reflect the affiliations of those involved – hence the scope for tensions and contradictions. This also means that responsibility for actions is diffuse and difficult to pin down – an important dimension given the legal and financial threats that more formal groups can face when defying the law (Wall, 1999; Rootes, 1997).

A key feature of many anti-roads protests was their generally local character, contrasting with the mainly national demonstrations that had characterised environmental and other protests during previous decades. That said, these local protests were designed to highlight perceived shortcomings in *national* policy – EF!ers opposed the roadbuilding programme as a whole, and wanted it to be reversed. However, participants were rarely naïve enough to believe they could reverse plans already being carried out. Rather, their aims were to slow down the process, to persuade individual construction workers not to continue, to increase the costs of roadbuilding for contractors and hence the government, and to transform longer term environmental and transport policy.

Friends and Enemies of the Automobile

To what extent is it fair to describe the critiques of 'car culture', the environmental policies and the environmental actions that have been described here as 'anti-car' – a label often given to them by their detractors (e.g. Dunn, 1998; Williams, 2000)? This phrase has gained considerable currency in recent years, especially since the publication of Labour's Transport White Paper (DETR, 1998). It has accompanied an increasing sense among advocates of car-based mobility that the car is 'in the dock' (Williams, 2000). This reflects a perception that transport policy has become more constraining of car driving, and that such constraints are unfair. The result has been a backlash against the consensus that developed over the course of the 1990s in favour of sustainable transport. Instead of concerns over how to solve the problems of congestion and pollution, proponents of the car focus on the problem of trying to express a cultural and in many cases individual 'love of the automobile' (Sachs, 1992) in the face of policies for car restraint. Commentators stress the importance of the car culturally, economically and for the benefits it offers to individual freedom (Dunn, 1998; Williams, 2000).

A well-presented example of this position comes from Dunn, who argues – from a distinctly American perspective – that the car provides 'a solution to most Americans' transportation problems' (1998: 1). For Dunn, the car empowers people by providing them with greater choice and helping a wider range of people become mobile; this explains why the car is so popular. He argues that solutions to transportation problems such as congestion and pollution need, therefore, to avoid marginalising either the car itself or the majority of people who benefit from it. He sees the car as central to such solutions, which will need to include car-based components such as technological advances that limit the negative environmental impacts of motoring, alongside a facilitative policy structure that can encourage alternative modes – but these should not at the same time attack car use.

This perspective is underlined to a large extent by research in the UK into the transport options of those on low incomes. Lucas et al. (2001) point out that the same policy measures can often address simultaneously concerns about the environmental impacts of transport and social equity issues. For example the health of disadvantaged people is often compromised by the location of social housing near routes affected by heavy traffic. However, they argue that restraining car traffic must follow rather than precede the provision of adequate alternatives, and highlight the ways in which access to a car can significantly enhance the quality of life of those living in areas suffering from high rates of social exclusion. In locations where public transport is felt to be inadequate, a car can be especially important in allowing choices which are otherwise more restricted, such as access to a wider range of employment locations, educational and leisure facilities.

Interestingly, these authors were struck by the hostility that their respondents showed towards policies for restricting car use through increased costs – these policies were regarded as 'attacking the motorist' (ibid.: 29) and Lucas et al. argue that raising fuel taxes is likely to 'attack' disadvantaged groups more than others, and thus exacerbate disadvantage. Their respondents expected to cut back on other essentials in order to absorb rising transport costs rather than switch to other modes which were generally seen as inadequate and unsafe. This hostility to the kind of measures labelled elsewhere as 'anti-car' was the case even for those who did not currently own a car, reflecting a widespread *aspiration* towards car ownership among those who do not currently own one (Glaister and Graham 1996).

A more strident hostility towards car restraint measures can be found in the public statements of the Association of British Drivers (ABD), a highly vocal campaigning group which is the operating arm of the tellingly-named 'Pro-Motor'. Whilst the ABD probably represents only some tens – or at most hundreds – of members, what is interesting about this organisation is the way it articulates its concerns as representing those of all motorists. Its defining statement, as set out on its website, is that '[t]he motorist in the UK is under attack from all sides'[3]. The ABD sees motorists as highly taxed, and subject to what it regards as unfair accusations that motorists cause pollution, damage the countryside and threaten the safety of other road users. The website sets out to refute these accusations with factsheets and other resources, drawing on scientific reports, car industry data and so on. The ABD states its objectives as being to promote: (i) the individual right to decide which transport mode to use, free of government interference; and (ii) making car ownership affordable right across society. It claims that 'these freedoms, which we have taken for granted for so long, are now under threat from a barrage of anti-motorist measures'. For example, it believes the 1998 White Paper was:

> drafted by anti car pressure groups and intellectuals, vested interests in public transport and certain commercial organisations who are too short sighted to see that clearing the roads for their companies will bring them little benefit if the whole economy suffers.

This perspective sees notions central to the White Paper such as integrated transport and modal shift away from car-dependence in somewhat conspiratorial terms that identify shared interests among a disparate range of positions – government, the commercial sector, even anarchists. In response to the riots in Westminster that followed an anti-capitalist demonstration in May 2000, the ABD identified similarities between the anarchist rioters and government policy. Its press release at the time read:

> These people are extremists ... But their philosophy can be found in every local transport plan. Their 'direct action' is balanced by other, superficially more respectable groups who quietly push within local and national government for anti car measures.

The sense within this organisation that motorists are under siege from every direction has been a common theme in recent transport debates. For example, the *Independent* newspaper reported in May 1998 that motoring organisations such as the AA and the RAC were concerned that 'environmentalists have dominated transport thinking'. Therefore, according to the newspaper, they were aiming 'to make a strong case for the car' in the run up to the publication of the transport White Paper, due two months later (*Independent* 30/5/98). Labour's integrated transport policy was duly labelled 'anti-car' by both the motoring organisations and the Conservative Party. The opposition transport spokesman suggested, for example, that '[t]he environment is just an excuse for Labour to fleece the motorist.'[4] This notion was repeated many times during the fuel crisis two years later.

The response of the Labour Government to this accusation highlights the complexity of the issue. Rather than there being a clear demarcation between, on the one hand, pro-car positions and the political right, and, on the other, anti-car positions and the left, the social benefits of the car, and the rhetorical power of car-centred discourse, cut across all sections of the political and social spectrum (Baird, 1998). The Labour Government has tried hard to deflect the anti-car label attached to it by others, by asserting its own car-friendly credentials, and by trying to extend debate towards broader transport issues. A leaked report in May 2000 stated 'Labour is not anti-car; we are anti-congestion; anti-pollution; and pro-safety' (*Independent* 31/5/00: 1), a position repeated many times in official transport statements. More interestingly, Labour appears to have revised its conception of how to achieve a more sustainable and integrated transport system. In December 1999, in the context of public questions over Deputy Prime Minister John Prescott's role concerning transport policy, Transport Minister Lord MacDonald was reported to have claimed that the Government no longer believed that cutting car ownership and use would reduce pollution. Rather, it now accepted the argument from car advocates that environmental problems could be solved technologically (*Daily Telegraph* 18/12/99: 1). Whilst this has not been borne out by subsequent Government policy, it indicates the way a variety of rhetorical strategies around these issues can be drawn on by political actors.

Just as Labour's position on the car has shifted around for essentially strategic reasons, so too have the Conservatives been keen to appeal to all sides. Their transport spokesman, Bernard Jenkin, again accused the Labour Party of being anti-car in October 1999, stating that 'The car represents real freedom and independence for ordinary people and Labour loathes it.' He portrayed his own party's position as 'not just pro-car ... We are pro-rail, pro-tube, pro-bus, pro-rural communities, pro-cyclist and pro-pedestrian.' (*Independent* 7/10/99). If nothing else, this statement illustrates how crucial politicians regard transport to be in securing support right across the electorate.

The difficulty of making the 'anti-car' label stick to policymakers brings us back to the earlier question of whether this position is a fair assessment of environmental campaigners. Even here, the evidence is uncertain. The strongest advocates of the anti-roads protests certainly identify themselves as vehemently anti-car. The article that launched EF!'s car campaign in 1991 decried 'the general level of apathy and hypocrisy of most green campaigns to date', in particular for accepting the notions of the 'responsible car driver' and the 'green car'. It argued:

> There can be no 'responsible car drivers' – when a car is driven, it pollutes and is inherently unsustainable; end of story. (Torrence, 1991)

Similar perspectives can be found throughout the literature of the radical environmental movement, and yet even here, condemnation is often tempered with acceptance of the everyday constraints on people's lives that make them dependent on their cars. Nicola Baird, for example, presents a highly critical analysis of the way cars are insinuated within economic and cultural life, yet opens her book with the line, 'This book does not blame car drivers for the state we're in' (1998: 11). This situation might, rather, be explained using concepts from within technology studies such as technological 'lock-in', technological 'obduracy' and the embeddedness of technology within social and cultural structures (see, for example, Sørensen 1994).

Resolving such multi-faceted problems is a major challenge requiring solutions that go beyond the purview of either pro- or anti-car campaigners. That said, the above discussion shows that neither the pro- nor the anti-position is uniform or wholly simplistic; both are mixed in how they articulate transport problems and solutions. A small (but very vocal) minority of environmentalists would proudly call themselves 'anti-car', whilst others both question the wisdom of having such a car-dependent society, but also accept the benefits – and sometimes the necessity – of car ownership. On the other side, the benefits of car ownership are trumpeted, and calls are made to end restrictions on car use. But at the same time, this does not necessarily imply hostility to other transport modes, and is, for some, part of an overall package that actively promotes other modes too.

These contradictory approaches to transport, the environment, the economy and the nature of our society underpinned much of the debate surrounding the fuel tax protests in September 2000. Far from being an entirely new issue that simply sprang up suddenly, the protests had many continuities with earlier developments. They followed closely from the 'Dump the Pumps' campaign that had failed to attract much public support just a few weeks earlier, but whose backers quickly became engaged in promoting the blockaders' cause[5]. Both these campaigns built on earlier work by organisations such as the AA and RAC in trying to publicise the fuel tax issue. In terms of discourse and rhetoric, the protests extended the critique of Labour transport policy discussed above, this time associating high fuel taxes with antagonism towards not just cars, but also small businesses, farming and rural ways of life.

These contrasts and continuities suggest some parallels between opposition to fuel taxes and opposition to roadbuilding. In both there have been a variety of different strategies, each appropriate to particular groups. More conventional organisations such as Friends of the Earth and the AA generally follow more 'legitimate' routes such as publicising the cause, and leave direct action to the less structured groups which have less to lose. With the latter, just as EF! and other environmental groups adopted 'repertoires of action' from previous campaigns, so too did the fuel tax protesters – often sharing strategies with the very same environmentalists whose positions they sought to counter. The next section will explore in more depth these parallels and differences between the two sets of protests, as part of an analysis of the different discourses that informed debate during the crisis.

Direct Action and the Discourse of the Fuel Tax Protests

An important dimension of the fuel tax crisis was the role played by the media in both shaping public debate and also disseminating information about the progress and effectiveness of the blockades. The way the media represented the crisis tells us a great deal about how these events were seen at the time, and also about how influential opinion-formers *wanted* them to be seen. One motoring organisation representative has commented, for example, on the way newspapers required a constant repackaging of information in order to extend the life of this 'story'[6]. It is instructive, therefore, to examine the ways in which transport issues during the fuel crisis were framed in the media, as an indication of how they were perceived more widely. The discussion in this section consequently draws primarily on press coverage of the crisis in order to explore the different discourses around transport that were deployed at the time. The aim is to assess not the historical accuracy of this coverage, but the ways in which the crisis was represented.

As has been shown with the strategies of both fuel tax and anti-roads protesters, public commentators often referred to earlier conflicts in assessing the tactics, official responses and outcomes of the fuel tax protests. In some quarters, the fuel protests were associated with earlier tax revolts (*Daily Telegraph* 14/9/00: 31), and even hailed (rather questionably) as part of a long tradition of 'great British revolts' including the Peasants' Revolt, the English Civil War, the suffragettes and the 1990 Poll Tax Riots (*The Sun* 12/9/00: 8). A planned convoy of lorries from Jarrow to London set to take place several weeks after the protests was compared by its leaders to the Jarrow March of 1936 – drawing an angry response from members of the labour movement who took offence at the comparison (*Guardian* 1/11/00: 11). This reflected the general position of trade unionists who recalled the very different ways in which striking miners had been treated by the government and the police in the mid-1980s. Unionists condemned the disruption caused by protesters against a democratically elected government by firms that would not hesitate to call in the police in the face of action by the labour movement (*Guardian* 14/9/00: 4, *Daily Telegraph* 14/9/00: 2). At the same time, it was pointed out that the fuel tax protests recalled:

> many traditional elements of trade union militancy – a feeling that politicians, whether they call themselves left or right or Third Way, don't really care; that the media distorts their case; and that employers or in this instance the government fails to negotiate unless pressure is applied through withholding services or labour. (*Guardian* 14/9/00: 5)

Despite these comparisons with older disputes, the more recent environmental actions described above provide the most pertinent parallel to the fuel tax protests. In their campaigning structures, the tactics they used and the topic on which they were focused, the fuel tax protesters bore a striking resemblance to groups such as EF! The media was fascinated by the fact that such effective protests were organised by a loose network of campaigns formed in recent years to highlight various aspects of the problems faced by farmers and small businesses. These included Farmers for Action, the People's Fuel Lobby, and the hauliers' group Trans-Action. It was also pointed out that those involved had close links with other right-wing campaigning groups. According to the *Daily Telegraph*, this protest constituted 'the Countryside March on wheels' (14/9/00: 31) – extending, therefore, earlier protests against what was perceived to be the primarily urban agenda of the New Labour Government.

Furthermore, although it was stressed that this was a spontaneous protest, many of those involved had also taken part in other actions, bringing into the fuel protests a range of previously-learned 'repertoires of action', just as had been the case with EF! Most notably, they drew on a similar range of modern communications technologies to mobilise support and publicity. The protests formed 'a network whose co-ordinating centre is constantly changing and linked by a myriad of mobile phones, CB radios, in-cab fax machines and laptops' (*Observer* 17/9/00: 16).

Where the fuel tax protesters differed from environmentalists was in the rationale behind their actions, and also in the scale and target of their objectives. As I have shown, environmentalist anti-roads protesters built their arguments on a range of issues concerned with alleviating pollution, congestion, ill-health and declining urban life, underpinned for many by a deeper critique of capitalist – and specifically industrial – society (Dobson, 1990). Whilst they aimed to delay individual development projects, they realised their best hope for effecting change lay in the influence their actions might have on longer term transport policy.

In contrast, the fuel tax protesters had their sights set on different goals – at one and the same time more narrow and more grand than the anti-roads protesters. Like the environmental 'movement', the apparently unified fuel tax protests actually masked a wide range of perspectives. At one level, this involved different approaches to the tactics adopted by the blockaders – how long should the blockades be maintained, to what extent was it safe to continue testing public support, which groups counted as 'essential users' in deciding to allow them continued access to fuel? Differences in approach within the network appear to have been seen as a weak point by the authorities – one newspaper reported that key groups had been infiltrated by MI5 in order to play up the differences between Brynle Williams and David Handley, two of the main players in the protests (*Observer* 5/11/00: 12).

More generally, there were a wide variety of motivations behind both the decision to take part in the blockades and the passive support given by a large proportion of the general public. The concern over the personal or commercial cost of fuel was predominant, but this overlapped also with concerns about the disproportionate effects of high fuel taxes on those based in rural communities and on poorer motorists. In addition, objections to the amount of tax levied on fuel blended in many accounts with a dislike of Government transport policies, and even of the Government *per se*.

This variety of concerns and perspectives meant that inevitably the protests and protesters were characterised by the media in a variety of often contradictory ways. The *Sun* newspaper's attempt to depict the protesters as heroic revolutionaries, mentioned earlier, indicates one end of the spectrum here. More credible was the *Daily Telegraph*'s portrayal of the protesters as people 'who want to be left alone by the state, and ask only to be able to make a living for themselves and their families' (14/9/00: 31). Such self-interest was a common characterisation of the protesters' motivations, and was interestingly given its most harsh expression again in the right-wing *Telegraph*, where one columnist referred to 'a protest motivated by hatred and greed' (Daily Telegraph 18/9/00:19). Individualism featured at least as much as the social dimensions of fuel costs in much of the media coverage of the protests, and it was matched in the public response to the crisis, panic buying petrol and food rather than seeking out more collectivist expressions of support (except for those members of the public who were local to the blockades). There is a striking contrast here with the anti-roads protests, whose motivations were primarily external rather than individual.

Yet the immediate goals of the fuel tax protesters were also far more ambitious than those of EF! The latter mounted high profile campaigns in order to gain publicity for environmental issues. A desire to harness the power of the media continues to be a central campaigning tool among such groups, as witnessed in the protests that have accompanied recent meetings of global economic interests in Seattle, London and elsewhere. In contrast, the fuel tax protests have aimed not so much to gain publicity but directly to bring about change, paralleling – as the quotation above shows – labour disputes with employers rather than environmentalists' disputes with government policy. Such a goal is something environmentalists have never imagined they could achieve, whatever their hopes might have been. In contrast, whilst the fuel tax protests were initially seen as a relatively localised dispute with only modest expectations of an impact, once public support – and public panic – grew, the protesters' demands for a result quickly scaled upwards. The result was their 60 day ultimatum to the government to meet their requirements or face further disruption.

Conclusion – Environment, Economy and Freedom in Transport Discourse

The timeliness of the fuel tax issue, in both political and populist terms, resembled to some degree the open door at which Greens had found themselves pushing in the late 1980s and early 1990s. For the Greens, this was quickly followed by a downturn in electoral support, in the face of national economic decline (Wall, 1999), but there was nevertheless considerable public backing for the anti-roads protests just a few years later. Yet this general concern about the environmental impacts of transport seemed, in the autumn of 2000, to have evaporated, a fate which subsequently befell the fuel tax protesters too as soon as their blockades were over – especially as the blockades had begun to impact on essential services.

Given the parallels, and the shared focus on transport, it is striking how little the environment featured as an issue for debate during the fuel crisis – especially since it was the Fuel Tax Escalator, an environmental measure aimed at cutting consumption in order to reduce emissions, that had sparked off complaints about fuel taxes during the preceding few years. It has been noted by many commentators that environmental arguments received very little airing during the protests, and that the Government appeared to be actively resisting putting forward the environmental case for the Fuel Tax Escalator. At the same time, only a few green voices were aired during the protests in the mainstream news media (e.g. Polly Toynbee in *The Guardian*, 15/9/00: 19; Jonathan Porritt in *The Observer*, 17/9/00: 29). The protesters therefore had little difficulty gaining the upper hand with their case for reduced taxes focused on economic arguments about fuel costs, the accessibility of transport and the lack of alternatives to the private car – thus ironically impinging on Labour's key policy area of social exclusion. The Government's response was framed primarily in terms of refusing to be forced into action by the protesters, rather than addressing the substantive issues of the debate.

Combined with other headline issues around transport at this time – most notably the crisis on the railways – the fuel tax crisis seems to have hardened motorists away from environmental considerations in what influences their transport choices. This situation is not in fact wholly new. A survey by the AA in 1999 found that the main issues of concern for drivers even then were the cost of fuel, the amount of fuel tax and the behaviour of other drivers. Concern about the impact of the car on the environment ranked lowest of all, with only 3 per cent citing it as their main concern, and only 15 per cent placing it in their top three (AA, 1999). The fuel crisis seems to have reinforced and perhaps exacerbated this situation. An RAC survey carried out during the September 2000 fuel protests found that the percentage of drivers willing to reduce their car use was at its lowest for eight years (RAC, 2001). Whilst this figure will have been influenced by the events taking place at the time, the RAC claims that it 'maintains a pattern which has been evolving since 1988'[7]. This survey also reiterates the empowerment conferred by car ownership as described earlier in the chapter (Dunn, 1998; Lucas et al. 2001). Sixty per cent of the drivers surveyed cited their sense of freedom as a key benefit they would miss without a car (RAC, 2001). The resonance during the fuel tax crisis of arguments about freedom and access – especially for rural drivers – was, therefore, not surprising, and this is something that has strongly characterised the discourse of pro-car advocacy (see Williams, 2000).

Yet the fragility of support for the protesters after the initial action had ended raises questions about the extent to which it is feasible to regard – and more importantly to mobilise – motorists as a distinct interest group. Discourses of choice and freedom – both the freedom to travel and freedom from regulation and taxation – provide a focus that distinguishes the political and rhetorical positions of groups such as the ABD and the Conservative Party from the kinds of objectives that have underpinned both environmentally-led transport policy and environmental transport activism. Younge and Glancey (1999), discussing the Conservatives' launch of their 'motorists' charter' in July 1999, pinpoint the potential that existed even then for capitalising on Labour's inability to gain public support for its transport policies. They write:

> scratch the surface and there is a powerful potential pitch for a new, classless society - the discontented driver ...

Their article continues – somewhat presciently – by arguing that this issue:

> has the potential of becoming a symbol in our political culture - a divider between public and private, between the rights of the individual against the intrusion of the nanny state. [Shadow Transport spokesman of the time] John Redwood uses terminology like access, opportunity, choices and taxes, not responsibility to the environment.

However, the very individualism that underpins this set of concerns also mitigates against the formation of an enduring coalition of motorists to pursue their cause. As Hallsworth and Tolley (2000) point out, the overconsumption of fuel that accelerated the crisis in its first few days can be seen as replicating the 'tragedy of the commons', that a mass pursuit of self-interest generally leads to a diminution of common resources. In this case, disregarding the needs of society more broadly resulted in the depletion of petrol stocks and the unintended consequence of bringing potential harm to essential services. The notion that all drivers share a common set of interests is convenient in the rhetoric of certain groups, but does not reflect 'reality'.

From an environmental perspective, the target of legislation and of much direct action should not, in any case, be drivers so much as the act of driving, as Harrington points out (in Williams, 2000: 74). Car ownership in itself is much less problematic (although not unproblematic) than car usage. This point tends to get lost in the polarisation of both pro- and anti-car rhetoric. Whilst for the former, the right to own and to use a car is presented as unchallengeable, for the latter the very existence of cars is objectionable. Such positions make it appear very difficult to find any common ground in what has become a highly contested area, yet they mask the great variety of perspectives that are covered within the broad categories of environmental and pro-car advocacy.

The polarisation of pro- and anti-car perspectives also masks the dynamic nature – and hence lack of stability – of this issue. This dynamism means that whilst transport seems certain to remain of central importance politically as well as environmentally for the foreseeable future, it is difficult to predict precisely which direction transport debates will take, nor what role transport activists can – or should – play in shaping transport policy. Whilst public and policymakers' attitudes towards transport issues have been shifting since the 1980s, so too have the positions of different interest groups, in ways that are highly strategic and not always consistent. Where politicians, activists and commentators stand often says more about agendas that are broader than just transport – taking in, for example, the wider urban-rural tensions that have been shown to characterise the relations between New Labour and many Conservative voters.

Perhaps, then, indeterminism is the key lesson to be learned from the shifts in focus that have taken place over the last decade or so. The move towards a greater environmental rationale behind transport policymaking turned out to be undermined by the embeddedness of the car within the everyday lives of ordinary people (Sørensen 1994). Whilst some seek ways of reversing this situation in order to overcome the negative effects of the car (e.g. Baird 1998), others argue that this is neither possible nor desirable (e.g. Dunn 1998). What is certain is that transport will continue to be an important political issue, and will continue to be contested, but that the ways in which it is framed, and the issues and practices debated, are bound to change over time. This means that the capacity of protesters to win over either the public or policymakers in future transport protests is likely to be highly context-dependent. As the fuel tax protesters found to their cost, what appeared in September 2000 to be the defining issue of the 1997 New Labour Government's period of office was no longer of much concern when it came to the General Election in June 2001. Widespread support around such an issue cannot, then, be taken for granted nine months, or even sixty days, later.

Acknowledgements

I wish to thank Brian Rappert, Sophie Taysom and three anonymous referees for comments on the first draft of this chapter, and David Williams of the AA for providing valuable background information. Parts of the chapter were presented within a conference paper, 'Hell is other cyclists: transport, policy and rhetoric', at the international conference, 'Worlds in transition: technoscience, citizenship and culture in the 21st century', Vienna, September 2000.

Notes

1 White Papers are UK Government publications that set out proposals for legislation.
2 The website for Reclaim the Streets includes coverage of anti-capitalism demonstrations as well as RTS itself. See http://www.gn.apc.org/rts/. Critical Mass can be found at http://www.critical-mass.org/.
3 All quotations from the ABD's website, at http://www.abd.org.uk.
4 Press release issued by Bernard Jenkin, 22/10/98.
5 See the Boycott the Pumps website at http://www.boycott-the-pumps.com/.
6 Private communication to the author.
7 Quoted from http://www.rac.co.uk/about/report_choices_1.

References

AA (1999). *The Great British Motorist 2000: Lessons from European Transport and Travel.* Basingstoke, AA.

Baird, N. (1998). *The Estate We're In: Who's Driving Car Culture?* London, Indigo.

Brass, E. and Coil, S.P. (1997). *Gathering Force: DIY Culture - Radical Action for Those Tired of Waiting.* London, The Big Issue.

Department of the Environment (1990). *This Common Inheritance: Britain's Environmental Strategy*, London, HMSO.

Department of the Environment (1996). *Planning Policy Guidance: Town Centres and Retail Developments (PPG6).* London, TSO.

Department of Transport (1989). *Roads for Prosperity.* London, HMSO.

Department of Transport (1996a). *Transport: The Way Forward.* London, HMSO.

Departments of Environment and Transport (1994). *Planning Policy Guidance: Transport (PPG 13).* London, HMSO.

Department of the Environment, Transport and the Regions (1998). *A New Deal For Transport: Better for Everyone.* Transport White Paper, London, TSO, and http://www.detr.gov.uk/itwp/paper/index.htm.

Dobson, A. (1990). *Green Political Thought: An Introduction.* London, Unwin Hyman.

Dunn, J.A. (1998). *Driving Forces: The Automobile, its Enemies and the Politics of Mobility.* Washington, D.C., Brookings Institution Press.

Elkin, T., McLaren, D. and Hillman, M. (1991). *Reviving the City: Towards Sustainable Urban Development.* London, Friends of the Earth/Policy Studies Institute.

Glaister, S. and Graham D. (1996). *Who Spends What on Motoring in the UK?* Basingstoke, AA Group Public Policy.

Green Party (1991). *Road to the Future: Global Warming, Congestion and Transport – the Solutions.* London, Green Party.

Hallsworth, A. and Tolley, R. (2000). Urban hysteria and panic in the streets - the British fuel blockade September 2000 - a social interpretation. in *World Transport Policy and Practice 6*: 5-10. http://www.ecoplan.org/wtpp/wt_index.htm

Hart, L. (1997). In defence of radical direct action: reflections on civil disobedience, sabotage and nonviolence, in J. Purkis and J. Bowen (eds), *Twentyfirst Century Anarchism: Unorthodox Ideas for a New Millennium.* London, Cassell.

Johnson, S.P. (1993). *The Earth Summit: The United Nations Conference on Environment and Development (UNCED).* London, Graham and Trotman/Martinus Nijhoff.

Lucas, K., Grosvenor, T. and Simpson, R. (2001). *Transport, the Environment and Social Exclusion.* York, Joseph Rowntree Foundation.

McKay, G. (1996). *Senseless Acts of Beauty: Cultures of Resistance Since the Sixties.* London, Verso.

Media Natura (1990). *Report on The 'Great Car Economy' versus The Quality of Life.* London, Greenpeace.

Melucci, A. (1989). *Nomads of the Present: Social Movements and Individual Needs in Contemporary Society.* London, Hutchinson Radius.

Owens, S. (1995). From "predict and provide" to "predict and prevent"? Pricing and planning in transport policy, *Transport Policy*, 2(1), 43-49.

RAC (2001). *RAC Report on Motoring 2001.* At http://www.rac.co.uk/about/report_intro.

Rootes, C. (1997). Environmental movements and green parties in western and eastern Europe in Redclift, M. and Woodgate, G. (eds.) *International Handbook of Environmental Sociology.* Cheltenham, Edward Elgar.

Roszak, T. (1971). *The Making of a Counter Culture: Reflections on the Technocratic Society and its Youthful Opposition.* London, Faber and Faber.

Royal Commission on Environmental Pollution (1995). *Transport and the Environment, Eighteenth Report.* Oxford, Oxford University Press.

Sachs, W. (1992). *For the Love of the Automobile: Looking Back Into the History of our Desires.* Berkeley, University of California Press.

SACTRA (Standing Advisory Committee on Trunk Road Assessment) (1994). *Trunk Roads and the Generation Of Traffic.* London, HMSO.

Sørensen, K.H. (ed.) (1994). *The Car and its Environment: The Past, Present and Future of the Motorcar in Europe. Proc. COST A4 Workshop.* Trondheim, Norway, May 6-8 1993.

Torrence, J. (1991). Carmageddon. Green Line 91 (October).

Wall, D. (1999). *Earth First! and the Anti-Roads Movement: Radical Environmentalism and Comparative Social Movements.* London, Routledge.

Williams, A. (2000). *Transport in the New Millennium.* Newcastle Upon Tyne: Transport Research GroupWorld Commission on Environment and Development (1987). *Our Common Future* (The Brundtland Report). Oxford, Oxford University Press.

Younge, G. and Glancey, J. (1999). Hot under the bonnet. *Guardian 13/7/99*, G2, 2-3.

3 Fuel Taxation and an Account of the Fuel Tax Protests in France

PATRICE SALINI

Introduction

The French road transport sector probably generates more blockades than any other sector. However, such behaviour is not unique to France (or indeed other European countries). Social protests use the blockade weapon to be heard. In France the road haulage sector, including both employees and employers, has known since 1984 that traffic embolism can be very effective when it wants to put pressure on the government. Sometimes, French blockades are instigated by Trade Unions, sometimes by Employers' Associations, sometimes by 'local committees' and sometimes by a combination of interests. The situation can be very complex. The most recent blockade, the fuel tax protest, was, as is often the case, the result of a combination of different strategies. These strategies are developed by Employers associations, which are in competition. Their oppositions - or alliances - are not only based upon ideological differences. They reflect different social and economic situations.

The French Fuel Protest

Before proceeding to explain the background and causes of the demonstrations of 2000, we need a factual account of the timing, nature and scale of the French protests.

According to the international media, Europe's fuel crisis started in late August with a blockade of English Channel ports by French fishermen. This protest spread across France as hauliers, farmers, and taxi owner-operators blocked oil refineries and depots.

Many analysts considered that concessions made by the French Prime Minister, Lionel Jospin, as a result of these protests were responsible for generating similar demonstrations in the UK, Belgium, Spain, the Netherlands and perhaps in Canada. In fact, the story is not as simple as that.

At the end of 1999, the Fédération Nationale des Transports Routiers (FNTR - the main French hauliers' organisation) pointed out the consequences for hauliers of fuel price increases and asked for a tariff revaluation. This first action clashed with a protest movement organised in February by French transport operators against the Aubry Law.

It is noted that, at that time, the International Road Transport Union (IRU) was:

> Hoping that, together, the French Government and the European Commission will succeed without further delay in rapidly restoring freedom of movement by making a clear choice in favour of a realistic and appropriate Community solution in the social field.

No-one spoke of fuel prices. In February, FNTR again voiced its opinions on a 'European commercial diesel oil', and continued with political action over fuel prices and transport tariffs.

At the beginning of July, the President of FNTR wrote to members to explain the timing of its different lines of approach:

- price negotiation with clients;
- government lobbying; and
- a likely demonstration in September.

During the summer, Mrs Loyola de Palacio was the first European Commission officer to acknowledge the disastrous effects that rising oil prices were having on the road transport sector.

The IRU, very pleased with Mrs Loyola de Palacio's initiative, called upon government in late August to:

> Take into consideration today's excessively high European fiscal charges that weigh heavily on the total cost of diesel.

According to FNTR and other operators' organisations, excises should be drastically cut.

Facing a lack of response from the government in France, FNTR, Unostra (another French haulage union) and Transports et Logistique de France (TLF) decided upon a blockade of oil refineries and petrol depots for 4 September.

The government fell short of expectations with a 'good but insufficient proposal' in response to the hauliers. It conceded a FF 0.35 per litre deduction in 2000 up to a maximum of 40,000 litres, a FF 0.25 repayment for 2001, and the suppression of excise rises planned by the government (as part of its climate change programme).

Only one organisation, TLF, signed a 'protocol' with the transport minister, Jean-Louis Gayssot and the blockade remained in place. On 5 September the Prime Minister announced that the government would act firmly against blockades.

The same day, FNTR called the government to engage in dialogue, which took place during the night of 7 September. Progressively, during the following night, FNTR trucks left the blockade points.

The end of the demonstration was, as is often the case, quite difficult for hauliers' organisations. Some of the regional and local presidents of FNTR disagreed with the protocol signed by the national federation. TLF left the demonstration following the first government proposals but Unostra did not resolve its decision to retire from the demonstration until one or two days following the decision by FNTR.

There were three main outcomes from the demonstrations:

- commercial tax repayment;
- a fiscal system based upon a tax damping when prices rise or fall; and
- global communication and fiscal policy in order to help the road transport industry to adjust prices to costs.

The first of these outcomes was temporary. The third outcome is more symbolic than effective. The second outcome is probably more significant, but less popular! These outcomes sparked similar protests to that in France by hauliers elsewhere in Europe.

The protests were similar in that they blocked highways, oil refineries and depots, clogged cities and created a fuel shortage. They were similar as well because the number of trucks required was not very large. In France, FNTR sought 4000 lorries to join its protest on 5 September. Only 1500 took part but this was sufficient, nevertheless, to achieve the desired effect.

French Road Haulage Industry

The French road haulage industry has a 'dual' image, both in France and abroad. On the one hand, there are some large companies: global European operators like Geodis, Norbert Dentressangle, etc. On the other hand, statistical data shows that there are more than 38,000 road haulage enterprises (employing 276,000 persons) of which about 12,000 are small businesses or owner operators.

In 15 years the sector has changed completely as a result of European integration and regulatory reforms. Before 1985, French rules were rigid, although less so than in Germany, and based upon a tonnage quota system and a mandatory regulation of road prices (Salini, 1980; Jones, 1995 and Neiertz, 1999). This situation, a quarter of a century year old, produced a protected sector, with a legal entry barrier, high prices and no internal growth. The new regulation system, introduced by the 'Loti' (Loi d'orientation des transports Intérieurs of 1983) and the end of price control in France in 1986, has generated a deep change (Bernadet, 1997). When rules change quickly, a new playing field opens, and a revolution in corporate strategies takes place.

This context was amplified by events in the European sphere and the failure to act against the European Council (Court Judgement of 22 May 1985[1]), the Single European Act and globalisation trends. In a few years, in France and in the major countries of Europe, the national context has been superseded and a new context has been set with common rules based upon non-discrimination and the free market.

This period was a revolution – a new era - for small enterprises, mostly established after the Second World War, in a context of strong regulation. They had to learn how to manage a 'normal' company without price regulation and protection. It was also a major upheaval (Salini, 1998) for the larger companies. It is well known that some of the major players, who were uncontested leaders twenty years ago, have now either vanished, dropped to the bottom of the league or quite simply been knocked off their perch. The French sector is no longer dominated by the large corporate firms and freight forwarders, who have been somewhat overtaken by the 'new' truck-load operators, integrators and subsidiaries of postal groups. In fact, the restructuring is a global one under the pressure of logistics outsourcing. Everybody wants to be global and be a third-party provider.

The survivors, both big companies and owner-operators, have produced a fantastic increase in productivity, largely shared with shippers. During the three to four years following 1986, road transport prices decreased by 15 to 20 per cent.

During the last period of twenty years before the end of the century, the internal outsourcing between truckers did not increase further, and was probably less important than under the tonnage quota and mandatory price regulation system. The image of a dual industry where freight providers, large companies, oppressed owner-operators or small sub-contractors, became somewhat outdated (Artous and Salini, 1997).

The new equilibrium, where national social laws were not well-respected, produced a success story for the industry. The only inconsistency, but a fundamental one, was social. The main conflict has been generated by a new driver's licence regulation in 1992.

The conflict was caused by a fight for hegemony between employers, workers associations and trade unions. The government changed its strategy and organised a forum designed to produce a real 'contract for progress' (contrat de progrès) between all the profession's players, both workers and employers, and the government (Dobias, 1994).

Two of the main issues were working time and wages levels. These issues were at the heart of strikes, blockades and conflicts between 1993 and 2000. The most significant strike happened in 1996.

> For the first time on this scale, workers began to come out from under the wing of the employers - said Claude Debons, General Secretary of CFDT-Transport, the most important drivers' association. (Hamelin et al., 1997)

It was one of the first steps along the new path: a real – not virtual – reduction of working hours. This process was very complicated and a continuous source of disputes, but it generated new social conventions and rules and a general climate of social negotiation. Companies tried to monitor and pay specifically for the hours worked instead of the usual lump sum. The next important step was to be the Aubry Laws and their implementation in the road transport sector, where working hours have a mean of 51 hours a week generally and 56 hours a week for long distance drivers. The objective was to obtain a realistic decrease of working time and not to align the road sector on the '35 hours a week' standard. Therefore the environmental context for the industry was changing, but in 1999 the French situation was not very different from the social situation in other European countries and from the French situation 25 years ago.

What is known in France as 'dualism' was not very important in social terms before 1993. There were no differences between working hours in large companies and small enterprises (Hamelin, 1997). Although the 'contrat de progrès' process began to generate a differentiation because of the trade union presence in large companies, the new situation was the same for everybody, and analysed as a threat to the industry. However, fortunately the fear was not considered great because of a high level of commercial activity. The consequences were not the same for the two parties, however. Large companies could outsource a part of their social expenses, whilst owner operators could not.

Many experts, both economists and sociologists, were surprised by the importance of the impact of the 'contrat de progrès' process on working hours. For them, the method, specifically based upon local negotiation and allowing that rules were not to be enforced until the end of the process, was very weak. In fact, surveys show that there is a real impact, but it is perhaps only a fragile impact.

Patrick Hamelin (Hamelin, 2000) explains that:

> Driving time decreases a lot for drivers who are not returning home for four nights or more per week.

For him, this change, which results in a decrease of the working time of long distance drivers,

> shows the nature of a changing relation with time because of the transformation of time accounting and reporting.

Table 3.1 Driving time and working time (hours) for long distance drivers (those not returning home for 4 nights or more per week)

	1983 survey	1993 survey	1999 survey
Driving time	39.7	42.6	39.3
Overall working time	60.8	60.8	56.1

Source: Inrets (Hamelin) surveys of 1983, 1993 and 1999

The dual impact was significant: the decrease was important for large companies and was slight for small enterprises and owner operators. A new differentiation appeared in France, which had not existed before. This difference still exists, for example, in the USA, essentially based upon the trade union action in some of the larger enterprises. It is probably the main indication of a social change in the French haulage industry.

But that was not the end of the story. Employers knew that the burden of wages would probably continue to increase while the sector suffered from a lack of drivers in a highly competitive environment. According to the 'Comité National Routier'[2], the theoretical labour cost for long distance drivers working the maximum 230 hours per month increased by 38 per cent between October 1995 and October 1999. Employers knew that a regulation would be published in early 2000, with new limits on working hours. They knew that 35 hour laws would generate a decrease of about 12 per cent of working hours.

Table 3.2 Working time in road freight transport enterprises after February 2000 (Décret n° 2000-69 du 27 janvier 2000 relatif à la durée du travail dans les entreprises de transport routier de marchandises)

	Maximum Working hours for a single week	Maximum working hours calculated for a month
Long-distance drivers	56 hours	50 hours per week, or 220 hours per month
Other workers (non-drivers)	48 hours	48 hours per week, or 208 hours per month

Representation and Competition

The economic and social background to the 2000 situation had become more complicated in terms of 'representation' and competition between employers' associations on the one hand, and trade unions on the other.

There are 3 major employers' organisations and 5 technical bodies dealing with social negotiation, worker and employer training, banking services, superannuation funds and economic observation.

The most important employers' organisation is the FNTR. It is considered to be the largest one in terms of member numbers (about 10,000) and probably in terms of road transport turnover.

FNTR views ultra-liberalism as dangerous for the sector and is asking for new commercial regulation. This regulation is not a return to the old system of quota and mandatory road price regulation, but it is difficult to imagine the type of regulation FNTR wishes coming into practice.

The basic analysis of FNTR is that margins are always decreasing although activity is rising and there is always a shortage of drivers. Prices are very inelastic and market forces are unbalanced. According to FNTR France does not have much clout in Europe.

For example FNTR wants to modify the legal basis of contracts, with an extension of the responsibilities of shippers and other clients and an increase in statutory pressure on 'low prices'. The main shippers' organisation (AUTF) disagrees with these views.

Another big organisation is TLF. This association is the result of a merger between freight forwarders, rental companies and carriers. The basis for this merger was ideological and economic. The largest companies are members of TLF and they have liberal positions. The number of members is probably about 1500, but represents a huge overall turnover.

This explains why they do not agree with FNTR demands. They use the same 'common negotiating body', UFT, to negotiate with trade unions. The most important enterprises, and especially the large groups, are members of both FNTR and TLF.

The last organisation is Unostra. The members (about 1000) are generally small enterprises. The history of Unostra is very interesting, because at the beginning, members were small, either communist or 'Poujadist'[3] employers. The result is a chaotic organisation, defending archaic claims, and trying to represent the smallest enterprises. Conflicts and strategic changes are sometimes very deep, but Unostra is often in competition with FNTR for leadership during conflicts with the government. It is evident that disappointed members of FNTR tend to join Unostra.

On the trade-union side there are 4 basic organisations. Three of them are federations of general trade unions like CGT (quite close to the communist party), CFDT, and CGT-FO. One specific organisation, FNCR, has both drivers and owner-operators as members. It is well known that the trade unions do not have many members in France, and especially not in the road transport sector.

There is competition too between workers' organisations. It is important to understand that transport federations have a special place in general trade unions because of the railway workers (who are numerous as members of associations), and the personality of some road federation leaders.

CFDT – the most important organisation for drivers – has chosen to promote and defend the 'contrat de progrès'. The others, due to a different combination of events and for political reasons chose not to support the process, or not to fight again. However, this deal took a long time to establish. One of the impacts of this process was an increase in trade union members, especially because of the importance of local negotiations.

What Happened in 2000?

The economic outlook was quite good for the road haulage industry both in 1999 and 2000. The level of metropolitan road transport, measured in tonne-kilometres, increased by 7.7 per cent in 1999 and by 2.6 per cent in 2000. The trend was very positive between 1998 and the beginning of 2000. It was very difficult to explain that the automatic effect of the increase of petrol was to slow the expansion rate of road transport. It was difficult too, to explain that the increase in the costs would have economic and political consequences.

The CNT (National Transportation Council) explained the risk at the beginning of 2000 in terms of a 'rolling-mill' of margins. At this moment we all know that the problem is not the excise level. The level of excise plays the role of a damping device, and the higher the excises are, the more important the cooling effect is. We know too, that, at this moment, the oil price is not a long term one, and it is possible to recognise that excises are not giving a good price signal.

This is not the main short-term problem. It is in the dynamics of the sector, which automatically produces a crisis. The global result of the year 2000 was not catastrophic, but the evolution during the year generated a crisis. Turnover increased by 6.7 per cent in volume in 1999, corresponding to an increase of 8.3 per cent in value and 1.4 per cent in price[4]. The estimation of the increase in turnover for 2000 was about 7 per cent in value with a 5.7 per cent price variation. It is not really crisis data. But the year's profile suggests a crisis.

There was a positive and significant continuous trend between the beginning of 1990 and the fourth quarter of 2000 in the cost level for road transport. This trend is evidently the most important in the last ten years and happened after a very slow trend during 1998.

Fuel was not the only factor. As the CNR (Comité National Routier[5]) explains, the impact on overall road costs of the increase in labour costs between January 1999 and July 2000 will be about 3.6 to 7.4 per cent (depending on the type of organisation and core business) in comparison with 5.4 per cent to 7.5 per cent for fuel costs. The 'cost wave' is significant and sudden.

Using a system dynamics model called 'Simtrans' (Karsky and Salini (1999) showed the changes during the two years 1999 and 2000 and helps to understand the reason why a blockade happened.

The model is based upon 'system dynamics', developed by Jay Forrester over the last 45 years. As Forrester says:

> System dynamics uses concepts drawn from the field of feedback control to organize available information into computer simulation models. A digital computer as a simulator, acting out the roles of the operating people in the real system, reveals the behavioral implications of the system that has been described in the model. (Forrester, 1991)

Simtrans is the first French transportation model based upon system dynamics.

The 1998-2002 profile is very interesting (see Figure 3.1). The model shows a continuous decrease of the margin rate during the year 2000. A delay appears between the changes in costs and in the margin rate.

Figure 3.2 shows the origin of this delay. It represents the trends of prices (per tonne km) and costs (per tonne km).

It is clear that the smoothing and delay effects generate losses between the end of the first quarter of 1999 and the middle of 2000. The level of losses is not likely to be totally accurate, but what is important is the trend. It is shown that problems appeared for enterprises during the second quarter of 1999, although their accounts probably did not really show this 'on the spot' loss. They mainly discovered the situation with the 1999 balance (generally closed at the end of the year), or when their bank suggested they reduce their short term debts.

If we analyse the cost trend and the price trend in terms of an annual moving average, it is clear that the maximum losses occurred at the end of the first quarter of 2000 with beneficial price adjustments imminent.

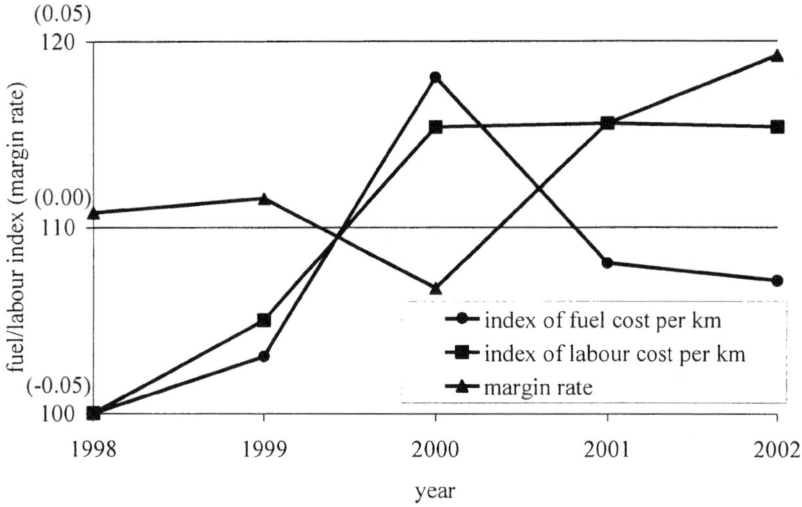

Figure 3.1 Simtrans simulation of the margin rate, labour and fuel costs (per km) – base 100 in 1998 for indices (labour cost and fuel cost)

The mechanism of economic delays is well known. When some commodities or their raw materials are more expensive, enterprises do not change their selling prices immediately. There is a 'delay', and probably some smoothing processes. These processes depend on the commercial relationships within the market. These relationships are not in favour of small truckers and sub-contractors. Another phenomenon has to be emphasised: the payment delays of shippers are probably more important than the hauliers' delay of payment for fuel and, obviously, for wages!

So, it is clear that a sudden increase in fuel costs or wages generates a temporary 'rolling-mill' of margins. The simulation (the model was designed before the oil price rise) demonstrates this.

As discussed earlier, this phenomenon is not the same for small enterprises or owner operators and the larger companies who have many sub-contractors.

On this basis we can understand the different strategies of employers' associations during the conflict with the government. FNTR and Unostra were clearly in competition to obtain the support of the majority of the enterprises. But TLF was satisfied earlier, when the government agreed to a reduction of taxes! This also explains the reason why the end of the conflict was so dramatic for FNTR, with an internal crisis with some of its 'departmental' organisations, which did not want to stop the blockade after negotiation with the Transport Minister. (FNTR lost members.)

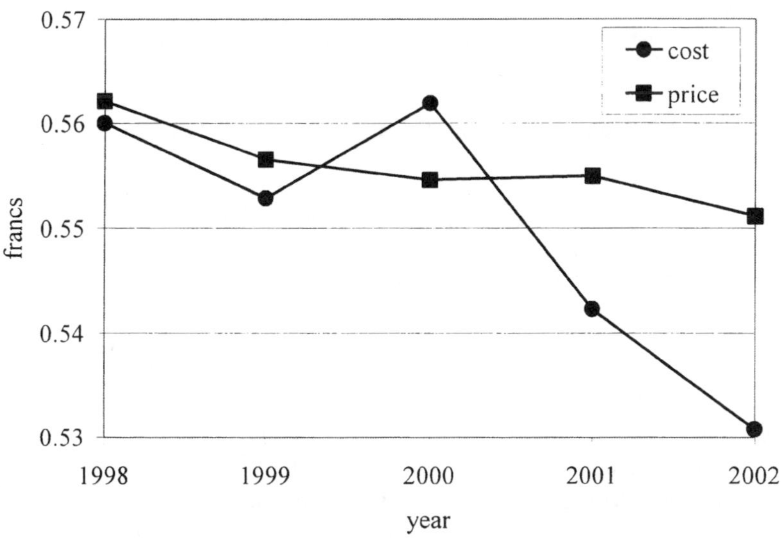

Figure 3.2 Simtrans simulation of sales and costs index (francs per tonne kilometre)

It is useful to read the explanations of the conflict by FNTR (FNTR's trucks represented 85 per cent of blockade vehicles) some months later. Guillemette de Fos, Communication Director of FNTR, April 2001, explained that:

> The conflict had been brewing for some months, and irritation was rising at the same rate petrol prices were increasing. In this context, the FNTR strategy was clear:
>
> Where, possible and for a long period, the organisation tried to help, the enterprises to reflect the increase of their raw material costs on their prices in terms of level and in terms of frequency. To reach this objective, the association published statistics, reports and so on.
>
> A second time during the spring finding out the continuous rise of expenses and the impossibility of continuing to claim to top up the prices, FNTR gave the alarm to people with press communications, and well documented synopsis were published, etc.
>
> Facing a global worldwide increase in the price of oil, FNTR decided to deal with the other employer organisations in order to "shout louder and farther". The aim was to take advantage of the French EU presidency to pass "a commercial fuel", that is to say a less taxed fuel in Europe. This action continued until the summer.
>
> Acknowledging as a fact that the response of the Ministry of Finance was not satisfactory, the principle of a blockade to take place on September 4th 2000 was decided at the end of August.
>
> The action was common to French organisations, but FNTR was not satisfied with the decisions announced on the second day of the conflict. These decisions were only dealing with an increase in excise recovery. It wanted – and obtained – decisions not only about expenses but also about receipts.
>
> FNTR obtained a study of an official costs index and new anti-dumping rules.
>
> According to FNTR, the conflict passed beyond the limits of the "Hexagon" as a testimony of the weakness of the haulage industry when expenditure rises are sudden and unexpected.
>
> Since that time, IRU has allowed this weakness in the sector to prevail. (Interview by the author with Guillemette de Fos, Communication Director of FNTR, April 2001)

As we can see, there was also a delay in the government comprehending the cost fluctuation problem. It is clear that at the beginning of 2000, the desire for new price regulations was not heard by the government. The French Transport Ministry had not noticed that the level of excises was an opportunity for governments to moderate the impact of a high level (or a low level) of the price of a barrel oil. The blockade is probably the consequence of a communication gap and an imprecise picture of transport economics.

The sector was not able to produce an adequate increase in productivity within a few months and was not able to find a short-term solution to a *sudden and unexpected* change. The structure of the sector, the transportation market and the number of small operators explains this inadequacy. It is difficult for the government to improve the sector based upon companies going bankrupt. It is much more difficult when the tax level is high under whatever pretext.

The structural problem and the weight of the political demands in terms of economic and social regulation are likely to induce further blockades. The new approach for organisations like FNTR is to demand new 'economic regulation'. FNTR explains that one of the ways of cleaning up the sector is the elimination of companies which pursue their activities through unfair competition.

The protests have generated a huge political gap between employers' organisations, and have put difficulties in the way of a sustainable policy.

For further background a CNR analysis titled 'fuel prices rise under control?' reads as follows:

> Any significant fluctuation in fuel prices causes an immediate effect on the profitability of enterprises. Explanations are simple. On the one hand, fuel costs are the second greatest expense after labour costs. For long distance transport, this expenditure may represent up to a quarter of overall operational costs. On the other hand, the cost variation repercussions are delayed and partial. Now, if the continuous decrease of fuel costs during 1998 allowed a marginal consolidation in the road transport sector, the sudden change of direction in the Brent quotation for 1999 January, under the effect of both oil price and Euro depreciation, destabilised lots of enterprises, in spite of tariff rises obtained from shippers.
>
> The average fuel cost (VAT excluded) moved from FF 3.18 per litre in December 1998 to FF 4.19 in December 1999, reaching a peak in September 1999 at FF 4.9, that is to say a rise of 50 per cent during the period.
>
> Considering the calamitous consequences of fuel price inflation, several European governments decided to lessen the 'cost shock', moderating the level of the specific fuel taxes.
>
> In France, the forecast rise of TIPP (French excise tax on fuel: Taxe Intérieure sur les Produits Pétroliers) has been postponed in 2000 (decrease in the excise of FF 0.2 in October) within the framework of a tax stabilisation system for all consumers along with an exceptional refund for some hauliers.
>
> This repayment moved from 3.54 centimes in 1999 to 35 centimes per litre (instead of 8.62 centimes previously). In 2001, this partial refund has been reduced to 25 centimes, but it may be completed by a tax damping whether prices rise or fall.
>
> This policy affects about 400,000 vehicles. It would cost 3.5 billion Francs over 2000. This is the reason why the government increased the oil refinery and distribution taxation payments in order to make these disbursements painless for tax payers.

Each month the Comité National Routier publishes reference costs index for general cargo. According to these results, without the government decisions the fuel expenditure would have increased by 24 per cent in 2000. These provisions limited the rise to 15 per cent. The smoothing effort has worked. But, we have to understand that this interventionism is necessarily temporary in a market economy. The agreements made on February 12 2001 corroborate this principle providing that the exoneration decided by Italy, Netherlands and France will cease at the beginning of 2003. In France, these repayments will be already deducted in 2002, the maximum being 40,000 litres a year instead of 50,000 in 2001.
(Special issue by Michel Hirou, Comité national routier 2001/4. CNT is a technical road transport organisation (one of the 5 technical bodies of the sector))

Notes

1 We read: 'On those grounds, The court hereby: declares that in breach of the treaty the council has failed to ensure freedom to provide services in the sphere of international transport and to lay down the conditions under which non-resident carriers may operate transport services in a member state'.
2 http://www.cnr.fr
3 Political doctrine of the 'Union de defence des commerçants et artisans' (UDCA) founded by Pierre Poujade in 1954. His doctrine was extreme right-wing, against parliamentary government and based upon specific claims.
4 Notes de synthèses du SES (2001) N° February-March.
5 CNR is a technical observatory tool of the employers' associations.

References

Artous, A. and Salini, P. (1997). Comprendre l'industrialisation du transport routier. Une modernisation contradictoire. Rueil, Editions liaisons, - 196 p., tabl., graph. Col: Transport logistique.

Bernadet, M. (1997). Le transport routier de marchandise fonctionnement et dysfonctionnements Editions Economica.

Dobias, G. (1994). Contrat de progrès pour le transport routier de marchandises Commissariat Général du Plan Rapport du groupe présidé par Georges Dobias.

Forrester, J. (1991). *System Dynamics and the lessons of 35 years*. Sloan School of Management, MIT.

Hamelin, P. (1987). Lorry drivers' time habits in work and their involvement in traffic accidents. *Ergonomics*, 30(9), London, Taylor and Francis, 1323.

Hamelin, P. (1992). Surveys about professional truck drivers. In *Selected readings in Transport Survey Methodology*, Melbourne, Eucalyptus Press.

Hamelin, P. (1997). Les conditions de travail et les carrières des conducteurs de poids lourds, analyse des résultats de l'enquête auprès des conducteurs de poids lourds, menée en 1993, comparaison avec ceux de 1983 ed. INRETS, Paris.

Hamelin, P., Le Coq, J., Deville, C., Ribeill, G., Debons, C., Artous, A. and Salini, P. (1997). *Les raisons de la colère*. ed. de l'Atelier, Paris.
Hamelin, P. (2000). La durée de travail des conducteurs professionnels comme enjeu de la flexibilité et de la compétitivité des transports routiers de marchandises et de voyageurs Working without limits? Re-organising work and reconsidering workers' health TUTB Parallel Workshops Rapport d'introduction.
Karsky, M. and Salini, P. (1999). Prospective des transports de marchandises en France à L'horizon 2020 Une application de la dynamique des systèmes complexes: le modèle Simtrans. Recherche aidée par le PREDIT.
Jones, J. (1995). Un siècle de politique des transports en France Editions Celse, Paris Loi n° 82-1153 du 30 décembre 1982 d'orientation des transports intérieurs modifiée.
Ministère de l'Equipement (2000). Circulaire no 2000-30 du 3 mai 2000 relative à la mise en œuvre de la réduction du temps de travail dans le secteur du transport routier de marchandises.
Neiertz, N. (1999). La coordination des transports en France de 1918 à nos jours. Comité pour l'histoire économique et financière, LGDJ diffusion, Paris, 798.
Notes de synthèses du SES (2001) N° February-March.
Salini, P. (1980). L'intervention réglementaire de l'Etat dans l'exploitation des transports terrestres de marchandises depuis la guerre. *Transports*, 256, 379.
Salini, P. (1998). Le grand chambardement. *Transport Magazine*, 157, 17.

4 Fuel Taxes and the Environment–Economy Trade Off

GREG MARSDEN

Introduction

In 1993, the UK Government introduced the 'fuel tax escalator', a measure to increase fuel prices in real terms every year. The increases were justified principally on environmental grounds as a means to help the UK reach its targets for reductions of carbon dioxide emissions and, through the reduction in mileage, to improve air quality.

Despite presenting the fuel duty escalator as an environmental measure, the Government only discussed the revenue-raising aspects of fuel duty during the fuel crisis of September 2000 and refrained from emphasising the environmental objectives of the measure. This chapter reviews the rationale for an environmentally linked fuel duty escalator through examining a range of evidence on the environmental damage caused by road traffic. The chapter addresses the question of whether the fuel duty escalator is intended to be an environmental tax or a revenue raising tax with environmental linkages.

The chapter begins with a brief discussion of the key environmental concerns faced by the Government and the policies they have put in place to manage these concerns. The fuel duty escalator has been the backbone of recent policy in this area. The need to consider costs of pollution in pricing transport is then explained to set the context for Government tax policy. Key statements from Budgets issued by the UK Government Treasury between 1993 and 2000 are then presented with a summary of the key changes to taxation policy and their environmental rationale. Next, a review of evidence from recent important studies on the unpaid or 'external' costs of transport is considered. The data are assessed to discuss the extent to which the current taxation policy does and could represent the full social costs of transport. Finally, the Government response during the fuel crisis is discussed before conclusions are drawn about the robustness of the policy approach currently taken. The chapter concentrates on the case with respect to car travel but identifies key issues and reports relating to the haulage industry.

Environmental Concerns

The UK Government's Guidance on Methodology for Multi-Modal Studies (GOMMMS - DETR, 2000a) lists the environmental factors that need to considered in the assessment of any proposed new multi-modal transport scheme as:

- local air quality;
- greenhouse gases;
- noise;
- landscape;
- townscape;
- biodiversity;
- heritage of historic resources;
- water environment;
- physical fitness; and
- journey ambience.

With regards to managing the environmental impact of existing transport networks, the traffic related impacts of the first three are most readily quantifiable and are currently being tackled through a variety of governmental programmes. This section briefly reviews the key impacts from these three areas and summarises the UK government policy to respond to these threats.

Local Air Quality

The Committee on the Medical Health Effects of Air Pollutants (COMEAP) advises the UK Government on the health impacts of air quality. In 1998, COMEAP submitted a report (COMEAP, 1998) concluding that the deaths of between 12000 and 24000 vulnerable[1] people may be brought forward by, and between 14,000 and 24,000 hospital admissions and re-admissions may be associated with, short-term exposure to pollution every year. Transport is responsible for significant proportions of pollutants emitted into the atmosphere, particularly in urban areas as shown in Table 4.1.

Table 4.1 Contribution of road transport to UK air pollutant emissions in 1997 (CVTF, 1999)

Pollutant	1997 national emissions (ktonnes)	Contribution from road transport % of national emissions	% of emissions in London
Benzene	38	65	82
1,3-Butadiene	10	77	97
Carbon Monoxide	5090	75	97
Lead	1.31	61	n/a
Nitrogen Oxides (NO_x)	1835	48	75
Particulates:			
PM_{10}	184	26	78
Black Smoke	314	59	n/a
Sulphur Dioxide	1660	2	23
Volatile Organic Compounds	2130	30	60

The Air Quality Strategy for England, Scotland, Wales and Northern Ireland (DETR et al., 2000) presents air quality objectives for seven pollutants for local air quality management and a further three objectives for national management. The air quality objectives dictate the number of times the national air quality standards can be exceeded per year. The standards are set at a level which 'can be taken to avoid significant risks to health'. The air quality objectives are however currently being breached at a number of sites in the UK, mainly for PM_{10} and NO_x. This has led to the forthcoming implementation of air quality management areas[2] – special zones with action plans to reduce pollution levels at a local level.

The solutions to the UK's air quality problems are a mixture of local and national initiatives across all sectors. For transport, in addition to local air quality management areas, the Government has brought forward a number of measures to reduce the environmental impact of transport, particularly financial measures to influence behaviour, summarised in Table 4.2.

Table 4.2 Fiscal measures from National Air Quality Strategy

Measure	Implication for air quality
Fuel duty escalator	Designed to encourage fuel efficient vehicles, reduce unnecessary journeys and make drivers consider alternatives to the car. A side effect will be air quality benefits if drivers cut down on short journeys in congested areas.
Fuel duty differentials	Significant way to encourage uptake of cleaner fuels (e.g. duty on ultra low sulphur diesel is 3p/litre less than on ordinary diesel and duty on road fuel gases is 1/5 of that on diesel)
Vehicle Excise Duty	Graduated Vehicle Excise Duty to be introduced based on vehicle CO_2 emissions and fuel type. Reductions also available for cleaner goods and public service vehicles.
Company car taxation reforms	Move away from vehicle value and distance driven related charges to system linked to vehicle value and CO_2 emissions.

Climate Change Gases

The 1997 United Nations Framework Convention on Climate Change in Kyoto agreed targets for industrial nations to reduce emissions of man-made climate change gases[3]. The European Union has agreed to an average reduction in levels of these gases to 8 per cent below 1990 values, over the period 2008-2012. The UK's individual target is a legally binding reduction of 12.5 per cent over the same period. The UK Government has also separately committed to a 20 per cent reduction in Carbon Dioxide (CO_2) emissions compared to 1990 levels by 2010 (DETR, 2000b).

In the UK, the transport sector is the fastest growing and third largest contributor to climate change gas emissions (Figure 4.1). Of transport greenhouse gas emissions 85 per cent comes from road transport (DETR, 1999) and, of this, 96 per cent is CO_2 produced through the combustion of fuel.

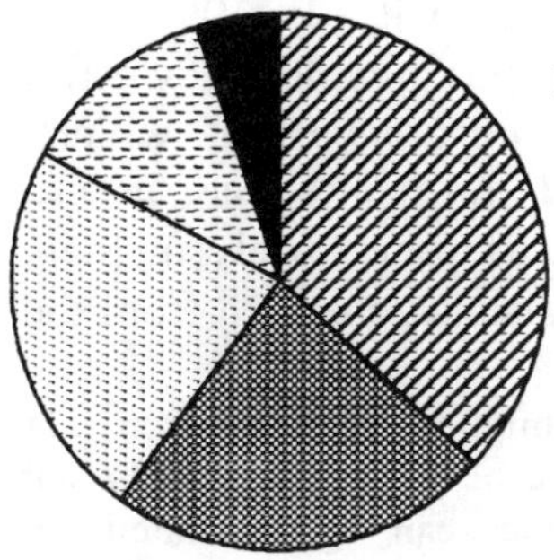

Figure 4.1 Climate change gas emissions by sector (from DETR, 1999)

In November 2000, the UK Government published its Climate Change Strategy (DETR, 2000b). The strategy sets out the Government's policies to manage climate change and provides a detailed policy breakdown to reduce emissions from each sector.

In 1990 greenhouse gas emissions from transport in the UK were 39.7 MtC[4]. The Government has forecasted that without action this will rise to 47.8MtC by 2010. Three key policy measures have been proposed to reduce this rate of growth. These are shown with their potential carbon savings in Table 4.3.

Table 4.3 Main transport climate change policies and their impacts

Measure	Carbon savings (MtC) to 2010
Fuel duty escalator to 1999	1 – 2.5
Reductions in CO_2 emissions from cars due to agreements with major world car manufacturers	4
10 Year Plan for Transport (DETR, 2000)	1.6
Promoting Cleaner Vehicles	Unquantified

Noise

Noise from transport is a well established source of pollution which causes annoyance to those subjected to it. In 1996 the European Commission issued a Green Paper (EC, 1996) in which they estimated 20 per cent of EU citizens were exposed to noise levels at which most people become annoyed, sleep is disturbed and health may be at risk. The GOMMMS acknowledges that individuals vary widely in their response to the same level of noise, even if the noise is generated from the same source.

> However the average or community response from a large number of people exposed to the same source of noise is relatively stable and a community average degree of annoyance can be associated with long-term average noise exposure. (DETR, 2000a).

Government policies to reduce noise have concentrated on improvements in vehicle noise and improvements in road surfacing to reduce tyre-road surface noise. The fuel duty escalator also offers some benefit in reducing noise by discouraging unnecessary journeys by car.

This brief review of Governmental policies to control air pollution, climate change and noise pollution does not cover all of the different measures in place to try and reduce the environmental impacts of transport. The Government strategy is generally to promote a mixture of technological improvements to reduce emissions per kilometre driven combined with fiscal measures to reduce the number of kilometres driven. Clearly, the fuel duty escalator has played a central part in developing the fiscal strategy. It was also one of the key motivating factors behind the fuel blockades.

Externalities

> An externality arises whenever an individual's production or consumption decision directly affects the production or consumption of others other than through market prices. (Begg et al., 2000)

In a perfectly competitive economic market place, consumption of a good will continue until the benefit derived from consuming one extra unit of the good is equal to the cost of doing so. Beyond this point, where the cost exceeds the benefit derived, the consumer would be better off not consuming. However, transport is an area in which the perfectly competitive market place often does not exist – one cannot buy rights to clean air or to noise levels and there is also some uncertainty in attaching values to these elements. In the case of transport one person's decision (driving a car from A to B) takes no account of the costs (or benefits) to other people such as worse air quality and increased noise. Under such conditions, the individual will drive more than they would were they to compensate the other people affected by the worsened environmental conditions caused.

One solution to the problem of externalities is to re-organise the taxation system such that users are faced with the full social marginal costs of travel for every journey (incorporating environmental, congestion and other costs). This pre-supposes that one is able to accurately determine externality costs and define a new collection system. The difficulty and expense of doing this has so far prevented it from being seriously considered as a solution. The current taxation system of Vehicle Excise Duty and Fuel Tax are therefore left with the task of equating private marginal costs with social marginal costs.

The next section reviews the UK Government's policy on taxing transport since 1993 through the budget statements delivered each year.

Governmental Fuel Tax Policy

Fuel duty has, since the early 1900s been an important source of Government revenue. Despite a series of ad hoc increases in fuel taxation throughout the 1970s and 1980s, the overall cost of motoring (of which fuel forms about 25 per cent (DETR, 2000c)) was falling. In 1993, the then Conservative Chancellor Norman Lamont instigated the now infamous Fuel Duty Escalator. In 1993, the duty was set to increase by 3 per cent in real terms each year. In 1994, Kenneth Clarke, the new Conservative Chancellor, extended this to 5 per cent.

> It is an essential part of the plans that I set out last year to deliver healthy public finances as quickly as possible and forms an important part of the Government's strategy to return carbon dioxide emissions to their 1990 level in the year 2000. (HM Treasury, 1994)

The 1994 budget also marked the beginning of the use of fuel duty differentials to balance out the environmental differences between different fuels.

> In recent years there has been a small differential between the duty on diesel and unleaded petrol. This differential is difficult to justify in economic, health or environmental terms. I therefore propose to tax diesel at the same

> rate as unleaded petrol. This means an increase of about 3 pence a litre on diesel. (HM Treasury 1994)

Duty on road fuel gases was also frozen.

The policy of pricing of fuel according to environmental damage was further applied in 1995 when the tax on super-unleaded fuel was increased by an extra 4p a litre (over the 5 per cent real-term increase) to reflect the higher emission of pollutants such as benzene (a known carcinogen). Duties on liquid petroleum gas and compressed natural gas were reduced by 15 per cent to encourage take up. 1995 also marked the move towards using Vehicle Excise Duty (VED), differentiated according to environmental credentials.

> We expect emissions of most pollutants to fall over the next few years. But emissions of some pollutants remain at high levels. So the Government intends to look at ways of using vehicle excise duty to encourage low emission vehicles. (HM Treasury 1995)

The taxation measures and environmental reasoning behind the measures between 1996 and 2000 are summarised below in Table 4.4.

It is worth noting that the change of Government in 1997 brought a Labour administration to power who continued and indeed increased the fuel duty escalator until 2000 when the escalator was scrapped.

One of the key protester groups during the fuel refinery blockades was farmers. A specially pigmented diesel 'red diesel' can be used in farm vehicles subject to a duty of 3.13p per litre (Bolton et al., 2001). Whilst this is a significant reduction compared to normal diesel, the farmers argued that all other aspects of their business are affected by transport costs not eligible for this rebate. Haulage companies are subject to diesel duty at the same rate as car drivers.

For further understanding of how fuel prices affect vehicle and fuel use, the reader is referred to Chapter 5 (Goodwin, 2002). However, it is important to note that at the outset of the fuel duty escalator, the economic benefits took a profile at least as high as the carbon dioxide reduction benefits in the budget statement.

The use of fuel duty differentials between fuels also implies that the Government recognises the need to price according to environmental damage and has some internal estimates, which were used to define the 3 pence increase on diesel. Indeed, statements by the Conservative Government in 1996 and subsequently by the Labour administration in 1999 suggest that the full social costs of transport should be paid.

Table 4.4 Summary of car taxation changes from UK Budget Statements

Measure	Treasury comments
1996	
5 per cent increase in duty in real terms	These tax rises will encourage fuel efficiency and help control harmful pollution
Reduction in Ultra Low Sulphur Diesel tax by 1p per litre compared to ordinary diesel	Produces less particulates than ordinary diesel but is more expensive to produce
Road fuel gases duty reduced by 25 per cent	Encourage a switch to cleaner gas power
1997[5]	
6 per cent increase in fuel duty in real terms	To support environmental objectives
Reductions of up to £500 VED for low emission lorries and buses	To encourage cleaner fleets
1998	
6 per cent increase in fuel duty in real terms	Only with the use of an escalator can emission levels (CO_2) be reduced by 2010 towards our commitments
Extra 1p per litre increase on ordinary diesel	
VED on smallest and cleanest cars reduced by £50 (1000cc and below)	Transfer some costs from ownership to use
1999	
6 per cent increase in fuel duty in real terms	Key policy for reducing CO_2 emissions
Expansion of VED reduction to 1100cc and below	
Announcements on reforms to company car tax and VED arrangements	Both to be linked to CO_2 emissions and fuel type from April 2002
2000	
No increase in duty in real terms	Due to oil price rise from 23$US per barrel to $31US
Ultra Low Sulphur Petrol 1p cheaper than normal unleaded	To encourage uptake
VED reduction extended to 1200cc and below	Announcement of new scheme for April 2002

> I firmly believe that motorists should bear the full costs of driving – not only wear and tear and congestion on the roads, but also the wider environmental costs. (HM Treasury, 1996)

> Economic instruments have an important role to play in influencing travel choice. They can ensure that the price of transport reflects all costs, including environmental costs. (HM Treasury, 1999)

There can therefore be little doubt, that the UK Government would like to see its vehicle and fuel taxation policies as delivering an efficient pricing system which reflects all forms of external costs (environmental, congestion, accidents etc.). The rhetoric in the successive budget statements reflects the increasing priority given to the environment within Government policy and the only mention of the economic rationale of fuel tax was in 1994.

Estimating Environmental Externalities

A number of recent studies have attempted to quantify the external costs associated with the environmental impacts of transport (Watkiss et al., 2001, DeLucchi, 2000 and Bickel et al., 1997). This section describes briefly how the estimates are derived and presents the range of values found. The reader is pointed towards the reports themselves for a more technical discussion of externalities. The external cost values are compared to the relative prices paid by motorists driving petrol and diesel vehicles in the UK at the time of the crisis to determine whether the Government's taxation policy matches the relative environmental costs (assuming congestion and accident costs are the same for different vehicles making identical trips). A comparison is also made between environmental externalities and externalities for congestion and accidents.

Delucchi (2000, p139) sets out the following framework for the estimation of the external costs of vehicle related environmental damage:

1. posit some change in motor-vehicle use;
2. estimate the change in air, noise, or water emissions associated with the change in motor vehicle use;
3. estimate the change in air quality, noise exposure, or water quality associated with the change in emissions;
4. estimate the impacts of the change in air quality or water quality; and
5. value the impacts in dollars.

The key environmental effects which have a monetary valuation are:

- Effects of atmospheric effects on human health;
- Effects of atmospheric effects on buildings and crops (non-health);

- Effects of noise on amenity; and,
- Effects of greenhouse gases on climate change.
  (Watkiss et al., 2001, p 81)

There are different considerations for each of the different effects. Health effects[6] are mainly as a result of exposure to pollutants directly emitted[7] and are therefore more severe in urban areas, as is the case with noise. However, effects of greenhouse gases occur at a global scale and have the same value per tonne emitted anywhere in the UK. The effects on buildings and crops is also dependent on pollutant transfer and falls somewhere between the two extremes. The implication of this is that environmental externalities will take a higher value per kilometre in built up areas than in less dense areas.

Table 4.5 shows the marginal cost estimates of environmental externalities as an average figure for the UK. The figure is averaged for all vehicle types over all areas.

Table 4.5 Average externality figures for the UK (Sansom et al. 2001)

Environmental effect	Cost estimate (pence per km)	
	Low	High
Air pollution	0.34	1.70
Noise	0.02	0.78
Climate change	0.15	0.62

An analysis of the different combined environmental costs of driving in different areas was also conducted by Watkiss et al. (2001). This showed that the marginal cost per kilometre for cars on any road in Central London was above 2.5p per km, more than 0.7p more than central conurbations. Rural driving is below 0.7p per km for all roads. A comparison between the ratio of costs of driving in Central London and Central Conurbations compared to the UK average is shown in Table 4.6.

Watkiss et al. also provide estimates of the difference in environmental costs per kilometre for petrol and diesel engines of varying technologies. Estimates of these values are shown below in Table 4.7 for urban and rural areas respectively.

Table 4.6 Comparison of urban environmental externality costs to national average costs

Area	Ratio of costs to national average
Central London (Watkiss et al., 2001)	1.6
Central Conurbation UK (Watkiss et al., 2001)	1.1
Los Angeles (Delucchi, 2000)	5.2
All US Cities average (Delucchi, 2000)	1.4

Table 4.7 Comparison of externality costs for petrol and diesel engine cars

Technology[8]	Cost (p/km)
Urban	
Diesel Pre-Euro	3.4
Petrol Pre-Euro	1.7
Diesel Euro I	1.6
Petrol Euro I	1.2
Diesel Euro II	1.3
Petrol Euro II	1.1
Rural	
Diesel Pre-Euro	0.9
Petrol Pre-Euro	0.8
Diesel Euro I	0.4
Petrol Euro I	0.4
Diesel Euro II	0.3
Petrol Euro II	0.4

The principal reason for the difference between diesel and petrol cars in urban areas is the higher emissions of particulate matter and nitrogen oxides from diesel vehicles compared to petrol. These pollutants are most at risk of exceeding the air quality guidelines for the UK and can cause severe respiratory health effects (COMEAP, 1998). The continuing improvement in diesel engine technology means that these emissions are being reduced and the gap between petrol and diesel is narrowing. A recent report by the UK Government-sponsored Cleaner Vehicles Task Force (CVTF (1999)) estimates emission performances relative to Pre-Euro I engines for urban driving (see Table 4.8).

Despite the narrowing of the gap, externalities from diesel vehicles are still greater than those for petrol engine vehicles in urban areas. A comparison of the fuel costs for driving 6000 miles of three equivalent model petrol and diesel cars is shown below in Table 4.9 for combined urban and inter-urban driving. Diesel cars have a fuel cost which is 65 per cent of that of petrol cars despite the higher environmental externalities (air pollution, noise and climate change). It is reasonable to assume that the costs of other externalities such as congestion and accidents are independent of the type of fuel that is used. This suggests that there is still considerable work to do to correct the relative costs of the fuels.

Table 4.8 Relative emission performance of petrol and diesel cars*

Type of vehicle	Emission Standard	Carbon Monoxide	Hydro-carbons	Oxides of Nitrogen	Particulates
Petrol Car	Pre-Euro I	100	100	100	5
	Euro I	15	9	19	2
	Euro II	10	4	9	2
	Euro III	7	3	6	2
	Euro IV	4	2	3	2
Diesel Car	Pre-Euro I	7	10	43	100
	Euro I	4	4	29	55
	Euro II	3	3	21	31
	Euro III	2	2	13	20
	Euro IV	2	1	7	10

*Carbon Monoxide, Hydrocarbons and Oxides of Nitrogen are all compared to Pre-Euro I petrol engine figures (base-line 100). Particulates are compared to Pre-Euro I diesel figures (base-line 100)

Average fuel consumption from new cars in the UK has remained almost unchanged since the late 1980s (CVTF, 1999). Whilst some of this effect is due to vehicles becoming heavier and a tendency to purchase larger cars (Brink, 2001), it also implies that fuel costs will be similar for drivers of older and newer cars, although the environmental externalities will typically be very much higher for older cars.

Table 4.9 Fuel costs for driving 6000 miles (VCA, 2001)

Model	Engine	Fuel	Fuel Cost for 6000 miles (£)
Citroen Xsara	2.0i 16v	Petrol	725
	2.0 HDi	Diesel	412
Ford Focus	1.8i 16v	Petrol	579
	1.8 TCi	Diesel	428
Peugeot 106	1.6	Petrol	653
	1.5D	Diesel	410

Other Costs and Externalities

Before it is possible to examine whether the price paid for fuel is representative of the marginal costs of transport, it is necessary to include all of the other costs including congestion, infrastructure costs and vehicle operating costs. Clearly, the marginal external cost of transport is highly dependent on the prevailing traffic conditions. The delay a journey causes to all other traffic affected by the trip will be much higher for a congested urban network than a rural trunk road, for example.

Sansom et al. (2001) have estimated the marginal external costs of congestion (p/km) for different road classifications in the UK (see Table 4.10).

Comparing these figures with those regarding environmental externalities, it can be seen that congestion externalities are much higher, particularly during peak periods and in urban areas. De Vlieger et al. (2000) measured real-world fuel consumption under a variety of driving conditions. They estimate that driving in congested conditions can increase fuel consumption by 20 to 45 per cent compared to off-peak. The figures for peak and off-peak driving from the table above show that congestion costs increase by around 100 per cent. Fuel consumption, and hence fuel tax, cannot therefore be deemed to accurately represent this increase.

The marginal costs, averaged for the proportions of kilometres driven on different road types in Great Britain, as estimated by Sansom et al. (2001) are shown in Table 4.11.

The results clearly show that whilst environmental externalities are an important element of externalities, they are dominated by congestion. The question is now asked as to whether the pricing structure of transport in the UK results in the correct marginal cost to deliver an economically efficient amount of transport. The summary of UK Government revenues from Sansom et al. (2001) is shown in Table 4.12.

Table 4.10 Estimated marginal external costs of congestion (Sansom et al. 2001, p29)

Category	Proportion of vehicle km (%)	Low p/km (1998 prices)	High p/km (1998 prices)
Motorway	17	12.80	12.80
Major urban central peak	1	85.76	85.87
Major urban central off-peak	3	46.61	47.15
Major urban non central peak	4	22.89	24.26
Major urban non central off-peak	8	11.09	13.67
Other urban peak	7	4.58	8.38
Other urban off-peak	15	0.65	4.89
Rural trunk and principal	30	9.11	9.21
Rural other	16	1.32	2.92
Overall	100	9.71	11.16

Table 4.11 Great Britain marginal road costs (p/km 1998 prices)

Cost	Low		High	
	Pence/km	% of total	Pence/km	% of total
Infrastructure	0.42	3.4	0.54	3.2
Vehicle Operating Costs	0.87	7.1	0.87	5.1
Congestion	9.71	78.8	11.16	65.4
Accidents	0.82	6.7	1.40	8.2
Environmental*	0.51	4.1	3.10	18.2
Other	-0.01	-0.1	-0.01	-0.1
Total	12.32		17.05	

*Air pollution, climate change and noise

From the marginal cost perspective therefore, the ratio of revenues to costs is 0.5 for the low externality estimate and 0.36 for the high externality estimate. These figures are overestimates for congested areas (which also have higher environmental externalities) and underestimates for rural areas (despite the increased accident risk on some rural roads).

It is worth noting that the results presented in this section refer to marginal costs. Sansom et al. (2001) estimate that 99 per cent of total social costs are recovered in the high estimate scenario and 213 per cent of costs are recovered in the low cost scenario through fares, vehicle excise duty, fuel duty and VAT according to the traditional cost recovery method applied by the UK Government. They conclude that despite the cost recovery outcome, 'Transport charges would need to rise if charges are to be set on economic efficiency grounds' (Sansom et al., 2001, piv).

Table 4.12 Great Britain marginal revenues (p/km 1998 prices)

Cost	Pence/km	% of total
Fares (bus and coach)	0.84	13.6
Vehicle Excise Duty	0.14	2.3
Fuel Duty	4.42	71.6
VAT on fuel duty	0.77	12.5
Total	6.17	

This commentary has focussed on car travel. It is worth noting that the marginal external environmental costs for an articulated goods vehicle are at least an order of magnitude higher than for a car (Watkiss et al., 2001). It has not been possible within this chapter to include a full analysis of goods vehicle pricing. However, two important studies have recently been published relating to goods transport. The then Department of the Environment Transport and the Regions commissioned the Lorry Track and Environment Costs report (DETR, 2000d) which estimated that trucks approximately covered their costs as far as these could be calculated. Since that time however English, Welsh and Scottish railways (EWS, 2000) reported that costs were underestimated in a number of ways and that trucks over 33 tonnes in gross weight did not pay their costs.

In either case, the rebate in diesel duty of 3p/litre, which followed the fuel crisis (Bolton et al., 2001), has meant that to some extent, costs are not being paid in full (Potter et al., 2001). However, whilst the amount paid by hauliers is clearly an issue, it is the costs that they face relative to other European countries which was the most contentious issue during the crisis. Such comparisons are not easy to make, particularly by comparing pump price alone. Burnham (2001) and Mitchell and Dolun (2001) explore these issues in more detail.

Summary of Pricing

The previous section has highlighted that fuel tax is only a weak proxy for environmental externalities that are highly vehicle-dependent and partly dependent on where pollutants are released. The research has shown that diesel vehicles are cheaper per kilometre to run by about one-third yet have higher external costs, particularly in urban areas. It is clearly difficult to have a single tax for one fuel with a variety of users of different characteristics operating on different routes – diesel goods vehicles and diesel cars being a case in point.

The Vehicle Excise Duty scheme, paid yearly by each user, according to vehicle type has recently been used to correct to some extent for the differences between vehicle types and fuel types. However, the difference in VED between a petrol and diesel engine vehicle under the new scheme being introduced in 2001 could be at most £60 – a figure that falls short of addressing the fuel cost differentials highlighted in Table 4.9. Whilst VED has helped to ensure that the Government at least recovers the total costs of transport, the full tax burden does not equate marginal social costs to marginal prices and is therefore distortionary.

Overall, and particularly in urban areas, congestion is the most important external cost identified. Research into the effects of congestion on fuel consumption has shown that increases in fuel consumption from stop-start traffic fall very short of accounting for the extra costs of congestion. Fuel taxes cannot therefore be seen as an effective way of discouraging urban journeys. The Transport Act (HoC, 2000) has made provision for Local Authorities to establish congestion-charging schemes within their boundaries. This is clearly one solution that would help to redress this imbalance in costs. However, the analysis by Sansom et al. (2001) has shown that there is an imbalance in transport costs, which goes beyond central urban areas which needs to be addressed.

Government Response During the Fuel Crisis

Despite the strong environmental rhetoric that has accompanied every budget statement since the introduction of the fuel duty escalator in 1993, the Government did not put forward the environmental credentials of fuel tax during the crisis despite some criticism of this in the media (Hewitt, 2000 and Porritt, 2000). Indeed, the statement below from the Prime Minister during the crisis is telling.

> Though it is absolutely correct to say that in the first two years of Government we did raise the fuel duty, it was necessary to sort out the economy. Tony Blair (Hughes and Rawstone, 2000).

The Government was criticised by a special report of the House of Commons Environmental Audit Committee into the Pre-Budget report which followed the crisis. The Committee was:

> disappointed that the Government only highlighted the revenue-raising aspects of fuel duty during the recent crisis and neglected to make the link to its role in meeting environmental objectives. (HoC, 2001)

The Government was not alone in failing to present the environmental credentials of fuel tax with William Hague, the opposition leader 'conceding that environmental concerns no longer play a part in fuel tax policy.' (Watson, 2000). Indeed, an anonymous colleague of the former chancellor Kenneth Clarke claimed, according to an article in *The Times* newspaper that:

> Ken never believed that this had anything to do with any of that environmental stuff. We had a £46 billion budget deficit and it had to be sorted out. The environmental excuse was a bit of cover. (Prescott et al., 2000)

The Government response during the blockades was surprising, given the significant body of background work which they conduct in support of the environmental credentials of the changes to fuel tax and VED. However, the lack of environmental focus can perhaps be attributed to the Government strategy of resisting the protest on democratic grounds. However, in so doing, the Government has weakened the link between fuel taxation policy and environmental goals at a time when it could have been strengthened. In response to the criticism raised by the Environmental Audit Committee the Government 'agrees that there is a need for all advocates of environmental policies to make their case more forcefully.' (HoC, 2001).

The Government's response to fuel taxes was delivered in the November 2000 Pre-Budget Report. The budget gave reductions of an extra 2p per litre on Ultra Low Sulphur Petrol (ULSP), compared to normal unleaded, making the differential 3p/litre. This was matched by a 3p cut per litre on Ultra Low Sulphur Diesel (ULSD), which accounted for almost 100 per cent of the market at the time. Fuel duties were subsequently frozen until April 2002 (Bolton et al., 2001). These reductions clearly went some way to appease public opinion on fuel duties and were sufficient to prevent a repeat of the protests. Whilst the introduction of ULSP will allow vehicle manufacturers to reduce emissions through improved engine design, ULSD already had 100 per cent market share and the environmental grounds for the reduction are not clear.

Conclusions

It would be naïve and unrealistic to report on the discrepancy between an efficient pricing system for transport and the Government taxation policy without commenting on the wider rationale for taxation. Excise duty on fuels is charged based on consumption and was originally introduced as a tax on a luxury good. Indeed, the Government has restated that:

> the appropriate level of fuel duties will be set on a Budget-by-Budget basis, taking account of the Government's economic and social objectives as well as the UK's environmental commitments (Bolton et al., 2001)

reflecting the importance of the revenue to wider objectives. However, it could also be argued that the Government has created unreasonable expectations of its own fuel taxation policy with various statements suggesting that fiscal instruments were being used to represent the full costs of driving.

A tax on consumption cannot reasonably be expected to reflect the full range of external costs resulting from the consumption given the inhomogeneous fleet of vehicles which uses our roads today. This chapter has presented a range of evidence which shows that transport users do not currently face the right costing structure to make efficient decisions. The work has also drawn on other studies to show that fuel tax cannot be used as a strong proxy for environmental or congestion externalities.

Environmental externalities form a much smaller proportion of social costs than does congestion (between 4 and 18 per cent of external costs compared to between 65 and 78 per cent). The evidence presented shows that despite fuel duty differentials, the balance between petrol and diesel prices does not reflect the difference in environmental damage caused. Environmental externalities are not constant over space, and increase significantly in urban areas through primary air pollution and noise pollution, an issue that is not reflected by fuel duty.

Increases in fuel consumption brought about by congestion have been estimated to be between 20 to 45 per cent, (and perhaps less in newer vehicles) which fall well short of the 100 per cent increases in congestion externalities. Whilst the Government has introduced legislation for road user charging schemes, the Chair of the Commission for Integrated Transport has recently cast doubt over whether Local Authorities will introduce the schemes to the extent that the Government expects (LTT, 2001). The Government has also ruled out the possibility of inter-urban road user charging until at least 2010.

It is not possible to imagine that the current system of taxation will deliver an economically efficient pricing structure for transport. This chapter has also shown that there are differences between vehicle types and areas and the cost and complexity of establishing a system that could monitor this may exceed the benefits of doing so. However, it is worth looking to the future to see whether the system will be robust in the face of advancing technology. Agreements with European, American, Asian and Japanese manufacturers have been made to reduce CO_2 emissions for new cars by 25 per cent by 2012. In addition, new hybrid engine vehicles and fuel cell vehicles will be introduced with near zero pollution. The Honda Insight hybrid car already offers a fuel consumption of 83.1 miles per gallon (compared to a fleet average of around 36). Given the dominance of congestion effects in total external costs, it will be difficult to capture this in a tax based on consumption of fuel. There are also distributional effects, with the poor being least able to take advantage of these improvements in the short term, and revenue implications of the changing fleet composition. These issues are discussed in more detail in Chapter 13 (Parkhurst, 2002). The Government should certainly give consideration to how revenues can be collected to reflect costs and equity issues in the medium to longer term.

Whilst this chapter shows that the vehicle and fuel taxation system is distortionary, the fuel duty escalator has indeed helped deliver reductions in CO_2 emissions as part of the Climate Change strategy. During the fuel blockades however, the Government failed to deliver any clear environmental messages in defence of higher fuel prices and instead stated the economic benefits of the tax. Indeed the evidence in this chapter shows that the Government's current taxation policy on vehicles is best described as a tax with environmental linkages rather than an environmental tax. Given the inhomogeneous nature of our transport system and vehicle fleet under the current system of payment this is the most we can expect.

Notes

1 People who already suffer from severe respiratory or coronary problems.

2 http://www.aeat.co.uk/netcen/airqual/aqma/home.html.

3 Carbon dioxide, methane, nitrous oxide, hydrofluorocarbons, perfluorocarbons and sulphur hexafluoride.

4 Greenhouse gas emissions are expressed throughout as million tonnes of carbon equivalent (MtC). 3.7 tonnes of CO2 contains one tonne of carbon.
5 1997 is the first budget of the Labour Government, which was in power at the time of the fuel tax protests.
6 See National Air Quality Strategy (DETR et al., 2000) for details of the health effects.
7 Some pollutants are formed from secondary reactions.
8 The technology refers to the European emission limits for new vehicles (see CVTF, 1999 for further details).

References

Begg, D.K.H., Fischer, S. and Dornbusch, R. (2000). *Economics.* 6th Edition, London, McGraw-Hill.

Bickel, P., Schmid, S., Krewitt, W. and Friedrich, R. (1997). *External Costs of Transport in EXTERNE.* Contract JoS3-CT95-0004, IER Germany.

Bolton, P., Seely, A. and Hough, J. (2001). Road fuel prices and taxation. *Library Research Paper 01/52* House of Commons. 11th May 2001.

Brink, R.M.M.V.D. and Wee, B.V. (2001). Why has car-fleet specific fuel consumption not shown any decrease since 1990? Quantitative analysis of Dutch passenger car-fleet specific fuel consumption. *Transportation Research D*, 6, 75-93.

Burnham, J. (2001). European Comparison of Taxes on Car Ownership and Us – Review of the Report for the Scotland Office by Colin Buchanan and Partners. *Research Report*, London, Commission for Integrated Transport. Available at: http://www.cfit.gov.uk/research/scot0122/index.htm [accessed 20/09/01].

COMEAP (1998). *Quantification of the Effects of Air Pollution on Health in the UK.* Department of Health, London, The Stationary Office.

CVTF (1999). *The Environmental Impacts of Road Vehicles in Use.* London, Cleaner Vehicles Task Force.

Delucchi, M. A. (2000). Environmental externalities of motor-vehicle use in the US. *Journal Of Transport Economics And Policy*, 34(2): 135-68.

DETR (1999). *A better quality of life: A strategy for sustainable development for the UK.* London, May, Department of the Environment Transport and the Regions.

DETR (2000a). *Guidance on the Methodology for Multi-Modal Studies.* Volume 2. London, Department of the Environment, Transport and the Regions. Ref: 99 AILT 1079 (Vol 2).

DETR (2000b). *Climate Change The UK Programme.* London, November, Department of the Environment Transport and the Regions.

DETR (2000c). *Transport Statistics Great Britain 2000 Edition.* London, Department of the Environment, Transport and the Regions, The Stationary Office.

DETR (2000d). *Lorry Track and Environmental Costs.* Report for Department of the Environment, Transport and the Regions by NERA, AEA Technology and the Transport Research Laboratory. Available at: http://www.roads.dtlr.gov.uk/nerarep/ index.htm [accessed 20/9/01].

DETR et al. (2000). *The Air Quality Strategy for England, Scotland, Wales and Northern Ireland.* London, Department of the Environment, Transport and the Regions in partnership with the Scottish Executive, The National Assembly for Wales and the Department of the Environment in Northern Ireland. Cm 4548 SE 2000/3 NIA 7.

De Vlieger, I., Keukeleere, D.D. and Kretzschmar, J.G. (2000). Driving behaviour and congestion: environmental effects by passenger cars. *Atmospheric Environment*, 34, 4649-4655.

EWS (2000). *Lorry Track and Environmental Costs*. London, English, Welsh and Scottish Railway.

Goodwin, P. (2002). Are Fuel Prices Important? In Lyons, G. and Chatterjee, K. (eds) *Transport Lessons from the Fuel Tax Protests of 2000*, Aldershot, Ashgate.

Hewitt, C. (2000). Drivers should pay. Institute for Public Policy Research in *Analysis*. The Guardian. 14th September, London. p.21.

HM Treasury (1994). The Chancellor's Budget Statement 1994. Available at: http://www.hm-treasury.gov.uk/budget/1994/speech.html [accessed 20/9/01].

HM Treasury (1995). The Chancellor's Budget Statement 1995. Available at: http://www.hm-treasury.gov.uk/budget/1995/speech.html [accessed 20/9/01].

HM Treasury (1996). The Chancellor's Budget Statement 1996. Available at: http://www.hm-treasury.gov.uk/budget/1996/speech.html [accessed 20/9/01].

HM Treasury (1999). Financial Statement and Budget Report 1999 – Building a Fairer Society. Available at: http://www.hm-treasury.gov.uk/budget/1999/fsbr/29805.htm [accessed 20/9/01].

HoC (2000). *Transport Act 2000*: 373. London, House of Commons.

HoC (2001). Government Response to the Second Report from the Committee, Session 2000-01 on – The Pre-Budget Report 2000: Fuelling the Debate. Environmental Audit Select Committee. London, House of Commons.

Hughes, D. and Rawstone, T. (2000). Blair: I will not surrender. The Daily Mail, 12th September, North West edition. p.2.

LTT (2001). Plan's congestion targets unlikely to be met says Begg. Local Transport Today, 23rd August 2001, Issue 322, p.3. Landor Publishing.

Mitchell, J.V. and Dolun, M. (2001). *The Fuel Tax Protests in Europe 2000-2001*. Royal Institute of International Affairs, Devon, Plymbridge Distributors.

Parkhurst, G. (2002) The Top of the Escalator? In Lyons, G. and Chatterjee, K. (eds) *Transport Lessons from the Fuel Tax Protests of 2000*, Aldershot, Ashgate.

Porritt, J. (2000). Is this the way forward? In *Comment*, The Observer, 17th September. p.29.

Potter, S., Enoch, M. and Fergusson, M. (2001). *Fuel taxes and beyond*. London, World Wide Fund for Nature: 38.

Prescott, M., Chittenden, M., Smith, D., Grimston, J., Rufford, N., Trump, S., Black, E., Parslley, D. and Rayment, T. (2000). Panic Stations: When patience and the pumps ran out. *The Sunday Times*, 17th September p.16.

Sansom, T., Nash, C., Mackie, P., Shires, J. and Watkiss, P. (2001). Surface Transport Costs and Charges Great Britain 1998. Leeds, Institute for Transport Studies in association with AEA Technology.

VCA (2001). *New Car Fuel Consumption and Emission Figures*. Vehicle Certification Agency. Available at: http://www.vca.gov.uk/fcb.htm [accessed 20/9/01].

Watkiss, P., Dore, C., King, K. and Murrells, T. (2001). Annex B: Environmental Cost Estimation. In *Surface Transport Costs and Charges Great Britain 1998*, Sansom et al. (2001), Leeds, Institute for Transport Studies.

Watson, R. (2000). Protesters deride Tories' 3p off petrol. *The Times*, 21st September. p.4.

5 Are Fuel Prices Important?

PHIL GOODWIN

Introduction

This chapter considers two different measures of the importance of fuel prices - their role in political action, and their influence on travel behaviour. The argument is that we may have underestimated the sensitivity of fuel price in both contexts, albeit using the word 'sensitivity' in two different senses which connect in a rather complex manner.

There was, undoubtedly, much complex interwoven politics involved in the fuel price protests of 2000, some of which - perhaps most - was only indirectly related to transport, or not at all. The role of the fuel distribution companies at the early stages remains obscure. There was a convergence of unrelated issues - the countryside, European lorry competition, even fox hunting - in which fuel price became symbolic of a wider political process. Commentators who usually would have found themselves resolutely on the side of 'law and order' could not prevent themselves from taking pleasure at the embarrassment of a Government they disliked. These underlying complexities continue, but if there is a next time it may be on a completely different issue, and may not represent the same combination of groups. Thus it would be wrong to take the protests only at face value.

On the other hand, it would be going too far to say that fuel price was only a convenient political handle. There were some startling features of the responses. Gimson (2001), writing in the *Spectator*[1] under the explicitly political title 'Another Victory Will Defeat Tony Blair', argued:

> The fuel blockade...was celebrated as a sign that at long last the people were fighting back against their oppressors

Nor were 'the oppressors' confined to Government: anger was swiftly transferred to fuel companies, for example in the typical newspaper headline (this one being from the London *Metro* (4.4.2001):

> Motorists outrage at £2.7 bn profit for Shell.

Responses were not only ideological. The market also moved in to offer new services, which were attached to the controversy, for example Birkett (2001):

> Britain's biggest ferry operator and a leading oil company have teamed up to offer haulage firms 'fuel cruises' to France to fill their lorries with cut-price diesel.
>
> The move...is likely to add to the debate over high fuel costs and was described by the Freight Transport Association as 'a reflection of the farcical tax situation which currently exists between Britain and France'.

For this chapter, they most significant aspect of these features of the protests is not their legitimacy or their possible re-emergence, but their *unexpectedness*. Government, most transport professionals and commentators (including the author), the media, the participants and even those interest groups who subsequently said 'we told you so' were surprised. They simply were not prepared for the intensity, scale, speed and prominence of the actions.

We can say, at a broad, descriptive level, that transport professionals in some sense underestimated the sensitivity of fuel price as a trigger of political action - underestimated the discontent which had become bound up with fuel prices.

This raises an unexpected connection. For some years, the author has been having a gentle argument with DETR (now DTLR) officials about the correct value of the elasticity of traffic levels with respect to fuel prices - i.e. if you put fuel prices up (or down) by 10 per cent, how much does traffic volume go down (or up) as a result. There is technical agreement that the effect is not huge, but the difference between 'small' and 'very small' and 'none at all' turns out to have quite big effects on traffic forecasts, the calculation of induced traffic, and assessment of which policies will be successful.

This is one of those technical issues which has significant policy implications, and is resolved ultimately by empirical analysis and research, which continues productively and is now within sight of resolution. But the background is that for at least ten years, and possibly twenty, the Government resolutely based its traffic forecasts and policy assessments on assumptions that suggested there would be a very small impact of fuel prices on travel choices or traffic levels. Although explanation for this practice is beyond the scope of this chapter, it may be noted that over the same period there were similar arguments challenging the official view about the assumed low (sometimes zero) incidence of traffic induced by road construction, or suppression of traffic by reductions in road capacity: there seemed to be a pattern that policy intervention in general was thought to have little effect on traffic, which was perhaps consistent with a 'predict-and-provide' policy orientation.

But in connection with prices, there is an intriguing and possibly important coincidence. In political terms, the sensitivity of fuel price was underestimated. And in traffic forecasting also, the sensitivity of fuel price was underestimated. The question naturally arises - is there a connection between these two features?

The answer turns out to be more complicated than at first appears, and the connection, if any, is certainly not direct. But in tracing this argument through, there are some potentially useful insights for both transport policy and analytical methods.

The Economic Theory - Ambiguity About the Word 'Sensitive'

In 1995 the author and colleagues wrote a report, published by the RAC (1995), which developed the theory that car dependence was marked by dynamic processes in which car use, essentially starting on a voluntary, discretionary basis, encouraged activity patterns, land uses, and life styles over a number of years. These in turn converted the discretionary activity into one perceived as necessary: i.e. car dependence did not just 'exist', it 'grew'.

This raised the long established distinction, in economic theory, between 'luxury' and 'necessity', which depends essentially on two empirical observations - what happens to people's purchases of various goods as (a) their income changes, and (b) the price level changes.

As income increases, consumers tend to increase their consumption of some goods more than others. Those goods which increase more than proportionally with income are defined as *luxury goods*, whereas those that increase less than proportionally with income are defined as *necessary goods*. (This does have a correspondence to the non-technical use of these words: a necessity has to be consumed even by people with low incomes, whereas a luxury does not.) Thus whether a good is a luxury or a necessity in this sense can be measured by its income elasticity, i.e. the percentage change in the quantity of the good demanded resulting from a percentage change in income. An income elasticity above one implies that demand increases relatively faster than income, whereas an income elasticity below one indicates that demand increases more slowly than income.

Another indicator of the difference between luxury and necessity, in standard economic textbook treatment, is given by its price elasticity, i.e. the change in demand for a good resulting from a 1 per cent change in price. As a general rule of thumb, we can say that a low price elasticity will be typical of a necessity - consumption, not being discretionary, is little affected by price changes. Similarly, a high price elasticity is associated with goods for which consumption is voluntary and discretionary, which is typical of luxuries.

Like all rules of thumb, there are important exceptions to this approach. There are theoretical complications about the relationship between price and income effects. The most important classical example was the observation that potatoes were so necessary for very low income people that if the price went up, demand actually went up (because no income was left over for a wider diet). And - as will become important later in the argument - there is a special case where a low price elasticity may arise simply because the prices are so low (relative to income) that there is little effect on choice at all. And in all cases, elasticity will be influenced by the availability of adequate substitutes.

With these caveats in mind, we return to the general case: examination of income elasticities and price elasticities provides us with a test whether consumption of transport - in particular, travel by car - is a luxury or a necessity, broadly defined here as discretionary or constrained.

This usage is reflected in the use of measures of benefit and loss in transport appraisals, especially consumer surplus. As far as possible, this is taken as an indicator of the change in welfare (or happiness) of a consumer faced with, say, a price change, and is influenced by the value of the price elasticity - a low elasticity[2], for a necessity, has a large consumer surplus and a price reduction will bring great benefit. A high elasticity, for a luxury, has a smaller consumer surplus and a price reduction will only bring a small benefit. Note that in this framework, *larger* behavioural responses are associated with *smaller* changes in consumer surplus, and vice versa.

Income Effects

Against this background, we can consider data on the relationship between car use and income, based both on comparisons of richer and poorer people at a point in time, and comparisons over time as incomes change. Figure 5.1, produced by Dargay for RAC (1995), shows the relationship between car ownership, and car use (in passenger kms), with real disposable income over the period 1952 to 1992. All three variables are in logarithmic form, and the slope of the line joining the annual observations is the income elasticity. For reference, lines representing the slope of an income elasticity equal to one are also indicated in the diagram.

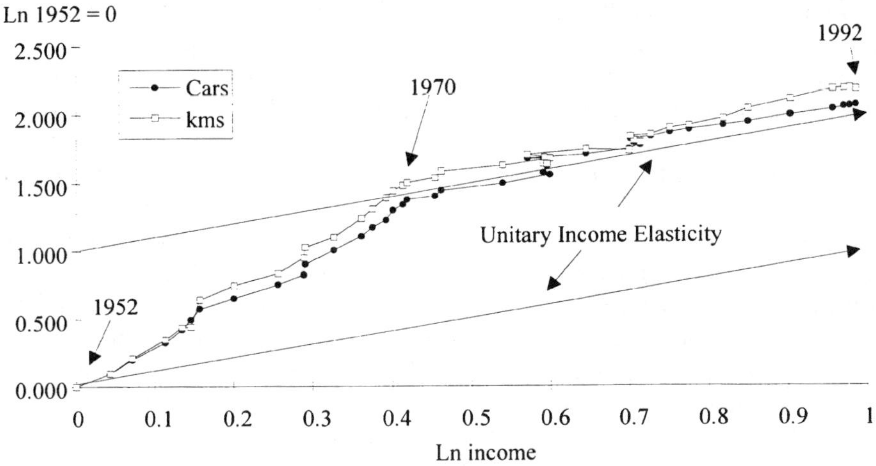

Figure 5.1 Cars and passenger kms related to income (Dargay, in RAC (1995))

It is apparent that up until the early seventies, both car ownership and use increased considerably more rapidly than income, indicating an income elasticity substantially in excess of unity. The growth in car ownership and use from 1952 to 1970 was over 8 per cent annually. If all this growth could be attributed to increases in income, which rose by 2.5 per cent per annum, the implied income elasticity would be around three.

Growth began to slow down in the early seventies, and the development in both car ownership and use more closely followed that of income. From 1970 to 1992, both car ownership and use rose by about 3 per cent per annum, while income increased by 2.5 per cent, indicating an income elasticity (again on the assumption that income is the only factor of importance) of about 1.2. The income elasticity remained above unity, so that car travel could be defined as a luxury good, but this was progressively less so: using the economic definitions, car travel was becoming less sensitive to income, but - by converting itself from a luxury towards a necessity - was becoming potentially *more* sensitive in terms of gain or loss in welfare.

Price Effects

The development of price elasticities shows a related, but more puzzling pattern.

In the UK, the early studies were influenced mostly by the work of Tanner, who for many years was the main person responsible for the UK forecasting practice (until his forecasting method was abandoned due to inaccuracies which do not derive primarily from its elasticities). Considerable attention was given to consistency between price elasticity and the generalised cost elasticities within which they sat. After a series of analyses in the 1970s, he estimated (Tanner 1981) a car ownership elasticity with respect to car purchase cost of -0.87, and with respect to fuel price of -0.31; also an elasticity of kilometres per car with respect to fuel price of -0.26. He interpreted these as long term elasticities, and used other evidence to suggest a *short*-term elasticity of around -0.15. He proposed a dynamic adjustment process such that the long term effect on car use was achieved over about 4 years (with the first year effect being about 40 per cent of the total), and about ten years for car ownership (first year less than 20 per cent of the total).

However, the 1980s were marked by a tendency for official traffic forecasts and policy appraisals to use smaller and smaller elasticities, culminating in the 1997 National Road Traffic Forecasts (DETR 1997, 1998) for which a figure of -0.13 was used for the long term effect of fuel cost on traffic levels. This followed a period, until revisions prompted by SACTRA (1994) when DETR also assumed that travel speeds had little or no effect on traffic levels - i.e. traffic was insensitive not only to price, but to most policy influence.

It was not at all clear at the time where these new, low elasticities had come from. It did not seem to correspond with wider evidence in the literature as a whole, in which there were some hundreds of studies - effects of fuel price on fuel consumption, and effects on traffic levels, and effects of road capacity changes. All these continued to find that traffic levels were influenced by price, speed and related variables. For example a review by Goodwin (1992) cited 13 studies in which the effect of fuel price on fuel consumption had been calculated, (giving a short-term elasticity of around -0.25 to -0.3 and a long-term elasticity of -0.7 to -0.8); 11 studies in which the effect of fuel price on traffic-levels had been calculated (with results of -0.16 for a short-term effect and about -0.3 for a long-term effect). SACTRA (1994) found that the effects of road capacity increases on induced traffic were broadly consistent with price effects at this level, and Cairns et al (1998) drew a similar conclusion for reductions in road capacity. Glaister and Graham (2000) again produced similar figures for fuel price elasticity from the research literature of the 1990s.

So although the orders of magnitude of elasticity found by Tanner in the 1970s, and by other authors in the 1980s, 1990s and early 2000s were rather similar, in the meantime the DETR used a series of revised figures whose effect was substantially smaller.

By early 2000 there seemed to be some degree of convergence on an agreed value, which was rather bigger than DETR had used in recent years, and closer to what it used some 20 years ago. This is shown in Table 5.1, as adapted from Goodwin (2001).

Table 5.1 Estimates of traffic sensitivity to fuel price

Source	Effect on traffic volume of a 10% price increase
Goodwin (1992)	-3%
Glaister and Graham (2000)	-3%
DETR (2000) input assumption	-2.3%
DETR model output for 2010	-0.65%

The difference between 2.3 per cent and 3 per cent, in this context, is a much smaller discrepancy than had previously been the case, and is probably to be explained by some differences in definitions, notably the difference between considering 'price' as per litre or per kilometre travelled, an important distinction generally but not material to the present argument. The important distinction, as it turned out, was that the future elasticity, as calculated from the model output, was very different.

Something was happening in the forecasts to suggest that, by the end of a decade, the sensitivity to fuel costs would be very low indeed - a ten per cent fuel price change would only cause less than 1 per cent change in traffic volume. The proposition is that, whatever the elasticity is now, it will be much less within ten years into the future.

Now bearing in mind the discussion above about luxuries and necessities, one possible interpretation of this could be that car use, and therefore petrol, is becoming more and more of a necessity, hence low elasticity, and this would tend to underpin the idea that changes in fuel tax, for example, would strike closer and closer to the heart of people's activities, and become more and more political sensitive. This would be the case if the forecast encapsulated an official view that car dependence would increase over the next 10 years, which would be very significant in view of the Government's intention to reduce it.

In fact the explanation seems to be quite the opposite. The reason for the forecast decline in elasticity does not fit *at all* into the interpretation of low elasticities as marking necessities or an increase in car dependence.

Rather, the DTLR's forecast that price elasticities will decline in future is driven primarily by its expectation that increases in income between now and 2010 would make people sufficiently rich that they would not really care so much about the money cost of travel - itself expected to get cheaper - which would therefore have little effect on their behaviour. The low elasticity in this case is a mark of fuel price simply becoming less important to behaviour, *not* a mark that travel is expected to become increasingly an economic necessity and less a luxury. Or in other words, prices simply will not matter very much.

The result raises a further, particularly important, aspect of the relationship between 'sensitivity' as expressed in elasticities, and 'sensitivity' as reflecting the importance to people's lifestyles. Thus we have two quite different reasons - with opposite interpretations - about what might follow from a tendency for the price elasticity of car use to decline.

The first interpretation (supported by some empirical analysis, but not used in DTLR forecasting) is that if the dominant trends of the post-war period continue, car dependence will increase, people will be captive to car, and they will not be able to adjust their behaviour to price changes. It would follow - with due caveats as for any political generalisation - that therefore any increase in price would be resisted more and more intensively.

The second interpretation (consistent with the generalised cost framework used in traffic forecasts, and used in the forecasting, but without strong direct empirical support) is that price will simply become less and less important, and the low elasticity would not be a source of greater political sensitivity, but less. It would follow - with the same caveats - that increases in price would be increasingly ignored as irrelevant and trivial.

Both of these interpretations are based on an expectation that price elasticities will in reality decrease over time as the forecasts suggest. This is itself not axiomatic, and no analysis has shown that a comparable trend has happened in the last 30 years, which, by the same assumptions, should have been the case. It is the author's view - though further research continues before this can be asserted with confidence - that fuel price elasticities will not decline substantially in the future, and therefore remain as a rather effective instrument for influencing traffic levels. But whether this, or the opposite, occurs, the same problem of interpretation will need to be resolved. It is clear that a high elasticity is associated with greater effectiveness as a policy instrument. It is *not* clear that political sensitivity is proportionate.

Conclusion

Politically, we have experienced an upheaval serious enough to constrain policy discussion, due, in part, to underestimating the 'sensitivity' of fuel price as a political issue. Analytically, we have a forecasting problem, affecting the way in which the transport policy and projects are evaluated, which may be due, in part, to underestimating the 'sensitivity' of demand to fuel price. The connection is, at best, complex. But both seem to indicate that policy discourse in the transport field has tended to play down the role of price, and we have two entirely different signals that it may be a serious mistake to do so.

There does seem to have been an underlying tendency for car use to have become more of a necessity, as perceived by drivers and as defined by economists. This has arisen for a number of reasons (including aspects not considered here such as public transport quality and land use patterns), and tackling this remains as a key task for public policy. However, there are some ambiguities in the evidence and interpretation which have significant policy implications. In particular, the empirical question - have the price elasticities been decreasing? - remains uncertain, and the interpretive question - if they have, does this represent a greater or a smaller political sensitivity of price? - remains open.

However, fuel price emerges as an influence both on travel choice and on political behaviour which is usefully open to analytical and empirical science. Rethinking our assumptions on this should usefully contribute to political and professional intuition, neither of which seems to have led us in quite the right direction.

Notes

1 A weekly journal known for its robust and consistent conservatism.

2 A caveat must be introduced here. Economists typically take an elasticity of 1 (i.e. proportional change) as being the bench mark for their descriptive language: a price elasticity of less than 1 is described as 'low'. This often causes confusion, because 'low' is interpreted to mean 'unimportant', or 'negligable' and hence by extension becomes transformed into 'zero' - a disastrously careless mistake. For practical purposes, a 'low' elasticity of say -0.3 might be very important indeed: if a fuel price decrease of 20 per cent causes an increase of 6 per cent in traffic volumes, the consequences for congestion and emissions would certainly be important.

References

Birkett, P. (2001). Stena runs fuel cruises for hauliers. *Evening Standard*, 4.6.2001, London.

Cairns, S., Hass-Klau, C., and Goodwin, P.B. (1998). *Effects of Road Capacity Reduction: Assessment of the Evidence.* London, Landor Publishing.

DETR (1997). *National Road Traffic Forecasts (Great Britain) 1997.* London, DETR.

DETR (1998). *National Road Traffic Forecasts (Great Britain) 1997*, Working Paper No 1, Car ownership: modelling and forecasting.

Gimson, A. (2001). Another victory will defeat Tony Blair. *Spectator*, 26.5.2001, London.

Glaister, S. and Graham, D. (2000). *The effect of fuel prices on motorists.* Hampshire, AA Motoring Policy Unit and UK Petroleum Industry Association.

Goodwin, P.B. (1992). A review of new demand elasticities with special reference to short and long run effects of price changes. *Journal of Transport Economics and Policy*, 26 (2).

Goodwin, P.B. (2001). *Running to stand still?* London, CPRE.

RAC (1995). *Car dependence.* P.B. Goodwin, editor, London, RAC Foundation for Motoring and the Environment.

Standing Advisory Committee on Trunk Road Assessment (SACTRA) (1994). *Trunk roads and the generation of traffic.* London, HMSO.

Tanner, J.C. (1981). *Methods of forecasting kilometres per car.* Report LR 968, Crowthorne, Transport and Road Research Laboratory.

PART II: EMPIRICAL EVIDENCE OF THE IMPACTS OF A FUEL SHORTAGE

6 The Effect of the Fuel Crisis on Sections of the English Motorway Network

DAVID EVES, JAMES QUICK, PAUL BOULTER AND JOHN HICKMAN

Introduction

TRL Limited analyse traffic data collected from selected sections of the English motorway network on a monthly basis. These data are taken from the inductive loop detectors which are an integral part of the motorway Incident Detection And Signing (MIDAS) system. The paired loop detectors are used to determine the flow, speed and length (categorised) of vehicles passing over them. Data are collected from each lane separately and loop detector sites are usually located at approximately 500-metre intervals. Data are reported at one-minute intervals, i.e. the number of vehicles passing over an individual loop pair and the average speed of vehicles in that minute. Categorised lengths are also reported each minute but are the aggregate of all lanes for a given site rather than individual lanes.

TRL conducted a study of the traffic conditions during the September 2000 fuel crisis and compared them with traffic conditions immediately prior to and afterwards. The analysis was based principally on MIDAS data. The intention of the study was to obtain a clear and unbiased quantification of the effects of the fuel crisis on sections of the motorway network, based on the available data.

Three sites were chosen from which to collect data, these were:

- M25, loopsite 4767, which is between Junctions 10 and 11 (to the south-west of London);
- M6, loopsite 5890, which is between Junctions 7 and 8 (near Birmingham); and

- M27, loopsite 9119, which is 500m downstream of the merge at Junction 3 (eastbound, near Southampton).

Both carriageways of the M25 and the M6 were examined. The M27 had only one carriageway (eastbound) examined. These motorways and sites were chosen to give a good geographical spread and the character of each was also known to be different. The M25 London orbital motorway is a very busy road with a high proportion of peak time traffic. The section of M25 studied is also congested, to a lesser extent, throughout the day. The western half also has a high density of junctions and several junctions with other motorways and major A-roads.

The site studied on the M6 is adjacent to the City of Birmingham and the East Midlands. The road serves both local traffic and traffic en-route between London, Bristol, Manchester and Liverpool. Peak hours are less evident on this section of heavily congested motorway, compared to the section of M25 studied.

The M27 is a relatively short section of motorway that links the adjacent sea ports of Southampton and Portsmouth and to the motorway network via the M3. It is a relatively uncongested motorway that shows distinctive increases in flows during morning and evening peak periods.

All the analysis concerned data collected between the 1st and 24th of September 2000 (weekdays and weekends). The blockades of fuel depots were between the 7th and 15th of September 2000, thus the base line performance of the motorways is illustrated along with any lingering effects.

The data were analysed to provide information on changes in flow, speed, the interaction between speed and flow, journey times, changes in vehicle composition and exhaust emissions.

Results

Total Daily Flows

Figure 6.1 shows total daily flows (i.e. 24 hours from midnight to midnight) for the M25. Figure 6.2 shows total daily flows for the M6. There were insufficient data from the M27, due to a loop fault.

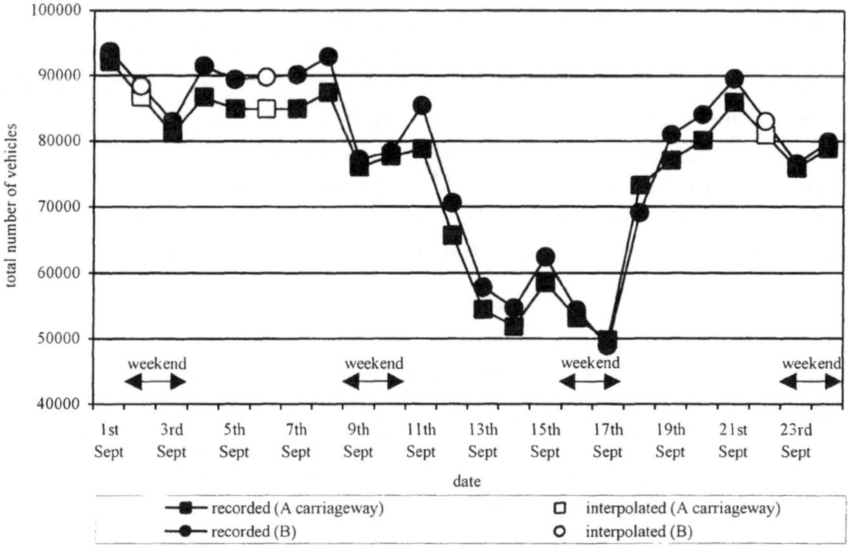

Figure 6.1 Total daily flows at loopsite 4767, M25 between junctions 10 and 11

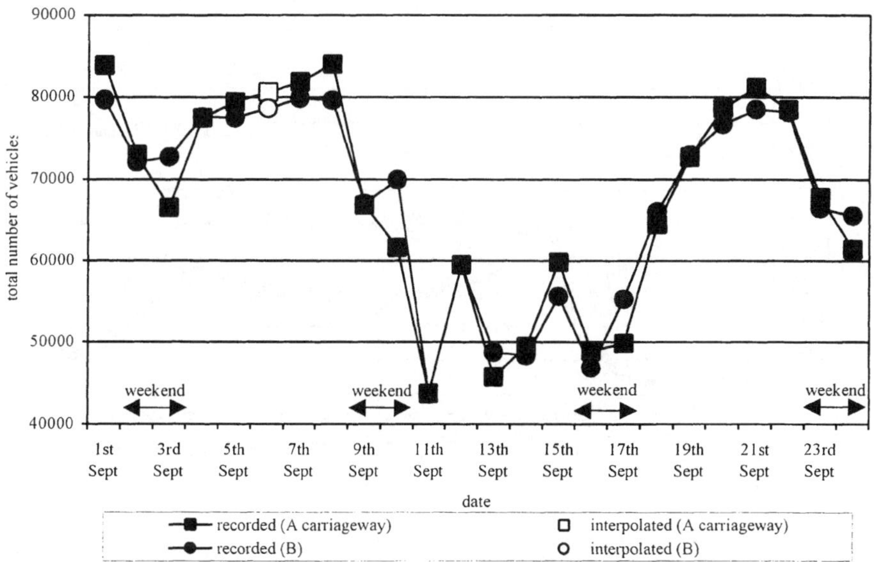

Figure 6.2 Total daily flows at loopsite 5890, M6 between junctions 7 and 8

The Figures show that there was a peak on Friday the 8th September 2000, the day after the first fuel depot blockade went in. The peak values are the same as for the same day the preceding week. The weekend then showed a characteristic drop in flow, however, this drop was greater than the preceding weekend. There was a slight rise for the M25 on Monday 11th September 2000, but then a severe drop after that for the rest of the week. Flows returned to normal the following week, on Wednesday 20th September 2000. There were a few days when there were missing data for all or part of the day. On these days, the values are linearly interpolated.

Daily Flow Profiles

Figure 6.3 to Figure 6.7 show flow profiles for the three motorways. A day in the preceding week (Thursday 7th September 2000) and a day during the fuel depot blockades (Thursday 14th September 2000) were selected for comparison. It should be noted that although the first blockade was on Thursday 7th September 2000, no substantial changes in flows were evident until the following Monday (11th September 2000). Thus, the 7th September 2000 was used to represent a before period. The flow profiles for the Thursday (21st September 2000) after the lifting of the fuel blockades were very similar to the profiles for the corresponding day before the blockades were imposed. They have not been shown in the graphs.

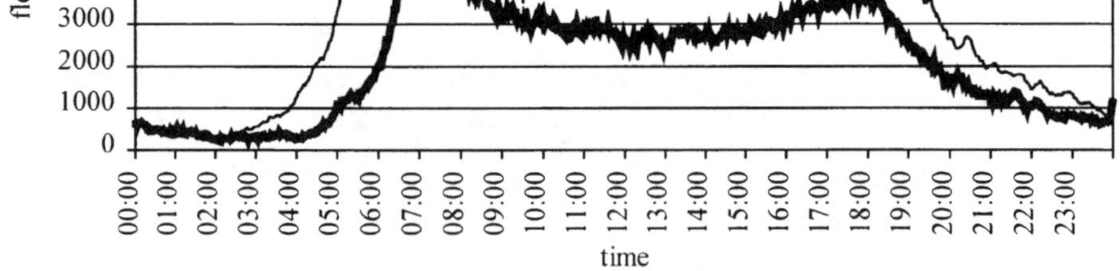

Figure 6.3 Flow profile for the M25 clockwise carriageway between junctions 10 and 11

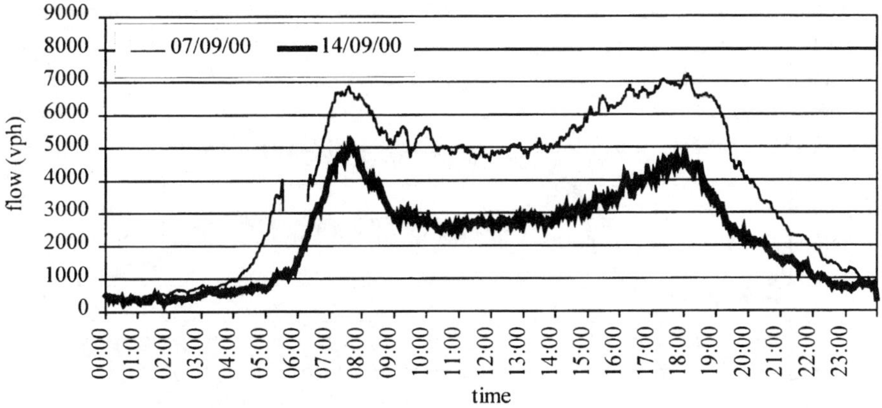

Figure 6.4 Flow profile for the M25 anti-clockwise between junctions 11 and 10

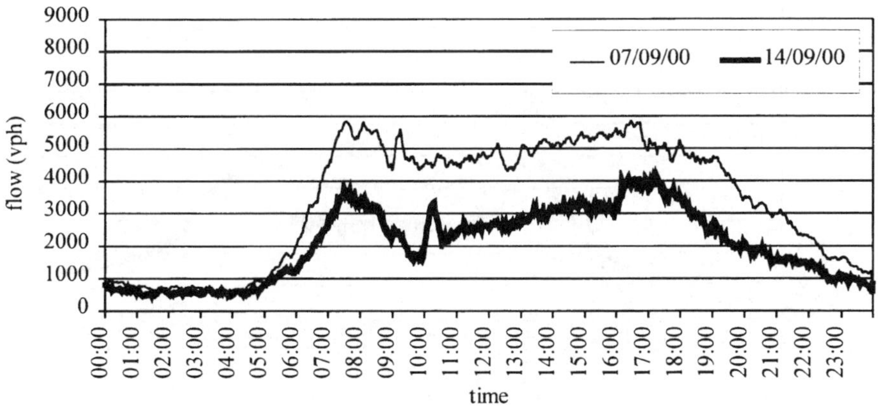

Figure 6.5 Flow profile for the M6 northbound carriageway between junctions 7 and 8

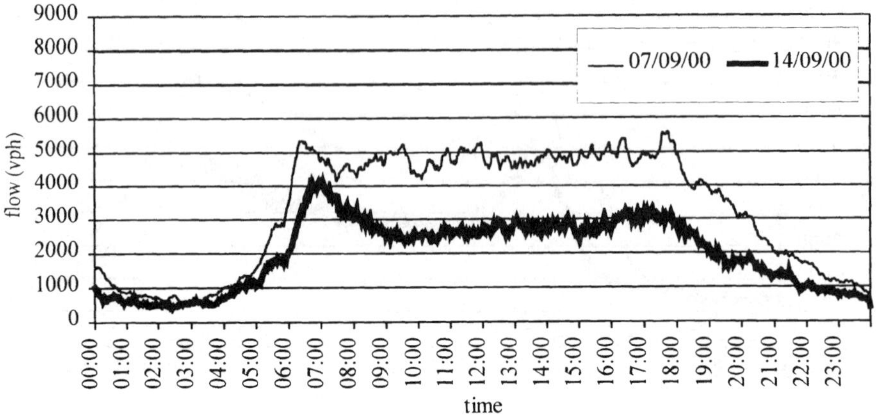

Figure 6.6 Flow profile for the M6 southbound carriageway between junctions 8 and 7

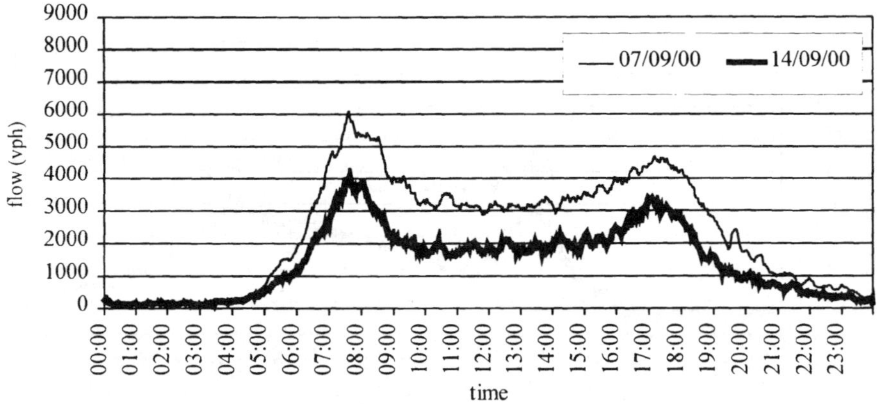

Figure 6.7 Flow profile for the M27 eastbound carriageway between junctions 3 and 2

The M25, for both the clockwise and anti-clockwise direction, showed typical flow behaviour before and during the fuel crisis ('typical' flow behaviour is characterised by a sharp peak in flow for the morning peak-period and an extended peak for the evening peak-period). In Figure 6.4 it can be seen that the time of the morning peak in the clockwise direction on the M25 did not change (07:30). However, the flows were much lower during the fuel crisis, approximately 5100 vehicles per hour (vph), compared with approximately 6600vph before, a reduction of about 23 per cent. Ordinarily, peak flows fluctuate (same weekday each week) by no more than 500vph, +/- 7 per cent (1 standard deviation = 280vph). At mid-day, the reduction was more pronounced. Again from Figure 6.4, the flows were about 2700vph during the fuel crisis compared with approximately 4800vph immediately before. This was a reduction of approximately 44 per cent.

The highest point of the evening peak occured at about 18:00. The timing of the peak did not change. Approximately 4500vph passed the site at the peak, compared with 6900vph on the selected day the week before, a reduction of about 35 per cent.

The M6 northbound carriageway showed typical, though fairly unpronounced, flow behaviour during the whole study period (Figure 6.5) with the period during the fuel crisis having less traffic. On the southbound carriageway (Figure 6.6) before the fuel crisis flow remained high between approximately 06:00 and 19:00, rather than showing typical flow behaviour. During the fuel crisis the flows were lower and showed more typical behaviour.

The M27 eastbound carriageway (Figure 6.7) showed typical flow behaviour for the whole period, with flows lower during the fuel crisis. The site on the M27 is relatively uncongested compared with the other two sites.

Note that the M6 and M27 have flows less than the M25. The M25 is a four lane site and the M6 and M27 are three lane sites.

Daily Speed Profiles

Figure 6.8 shows the speed profile for the clockwise carriageway of the M25 for the same days as depicted in the flow profiles. The daily speed profiles for the anti-clockwise carriageway of the M25 and the other sites show similar characteristics.

The speeds on the motorways were more consistent during the fuel crisis than the preceding week. There was no congestion and no flow breakdown (flow breakdown is generally characterised by 'stop-start' conditions and a reduction in flow due to congestion) during the morning and evening peak periods during the fuel crisis. There was also no normal congestion throughout the day on the M25 or M6 (the M27 does not normally experience congestion between the morning and evening peak period). Speeds during the peak periods at the M25 site were considerably higher than are normally experienced. Speeds were also marginally higher between the morning and evening peak periods. However, the maximum speeds (normally experienced during the night) were not as high during the fuel crisis as they were prior to it.

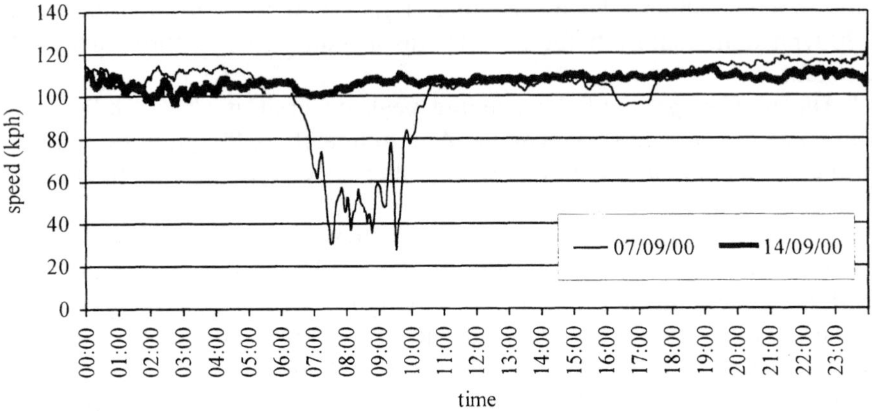

Figure 6.8 Speed profile for the M25 clockwise carriageway between junctions 10 and 11

The pattern was similar for the M6, with there being no flow breakdown experienced during the fuel crisis, although for the late night hours it experienced higher speeds than before the fuel crisis. The M25 pattern was also repeated for the M27, with there being no flow breakdown during the fuel crisis.

At the time of the fuel crisis, there was a perception that drivers were driving more slowly than normal, in order to conserve fuel. However, the above analysis shows that drivers actually drove more quickly than normal during daylight hours because of lower flows and the absence of congestion. However, a clue to support the perception above can be found during the night. At this time speeds were lower than normal and this is a time period when congestion is not normally experienced on the M25. This leads to the hypothesis that people's desired speed for a given flow level was lower during the fuel crisis.

Speed/Flow

To test this hypothesis, the average speeds corresponding to differing flow levels were plotted.

Figure 6.9 and Figure 6.10 show the speed/flow graph for the site on the M25 and the site on the M6 (the M27 site showed a similar pattern). The graphs show the average speed (kph) for each flow level (vph) on the 7th September 2000 and the 14th September 2000 for a single lane. For clarity, the speed/flow for after the fuel crisis is not displayed, however it is similar to the before period. As flow level increases, speed decreases until the point of flow breakdown occurs. At this point, a transition occurs and both speed and flow decrease (flow breakdown).

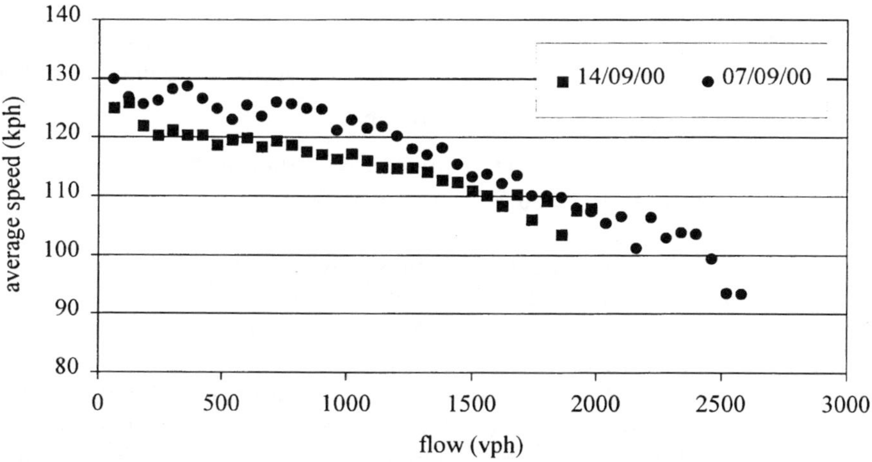

Figure 6.9 Flow/speed chart of the M25 clockwise carriageway between junctions 10 and 11 for the offside-1 lane

For clarity, all points with a speed less than 90kph were discounted. This is because there was no occurrence of congestion and flow breakdown on the 14th September 2000 so no comparisons could be made. The value of 90kph was determined after studying a speed/flow scatter graph displaying all the values and identifying the point at which flow breakdown was initiated.

Both graphs show that, for a given flow level, the corresponding speed chosen by drivers was lower on the 14th September than on the same day in the preceding week. At low flow levels, the differential between the desired speeds before and during the fuel crisis was greater than at higher flow levels. As flow levels increase, the speeds of vehicles before and during the fuel crisis converge. This is probably due to the gradual disappearance of the ability to travel at desired speeds caused by the onset of congestion.

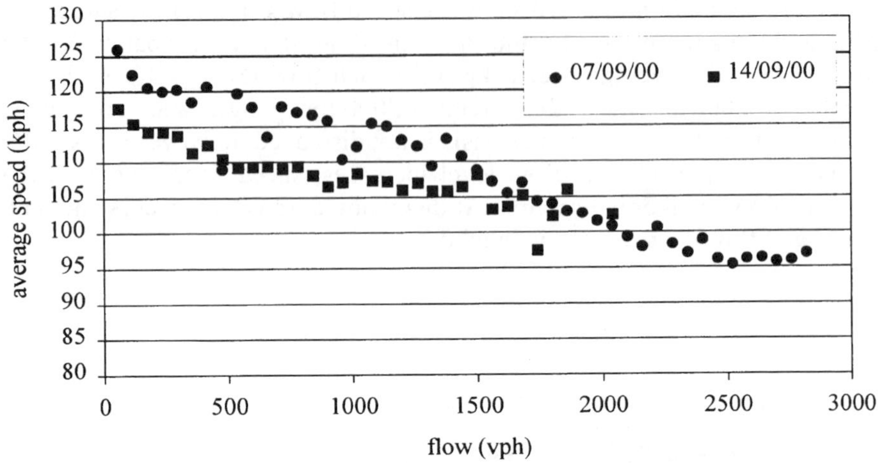

Figure 6.10 Flow/speed chart of the M6 northbound between junctions 7 and 8 for the offside lane

There are three major issues that may affect these speed/flow curves. First, there may be a bias due to different desired speeds for day and night. That is, before the fuel crisis low flow levels were only experienced at night, whilst during the fuel crisis, low flow levels were experienced during daylight hours as well. However, it would be reasonable to suppose that desired speeds, if different, would be lower at night than during the day. This would work to decrease the differential in speeds found. All the sites analysed are lit and it is known that, in such conditions, drivers' desired speeds are the same as for daylight conditions. Note that, despite higher speeds at night on the M6 during the fuel crisis, the speed/flow curve shown in Figure 6.10 clearly shows the differential in desired speed.

The second source of bias may be due to weather conditions. Adverse weather conditions would lead to a lower desired speed. However, weather conditions were good on both the 7th and 14th September 2000.

The third source of bias may be due to changes in vehicle composition during the fuel crisis, compared with before and afterwards. An increase in the proportion of heavy-duty vehicles could cause a drop in the average speed due to the lower speed limit imposed on these vehicles. However, the proportion of heavy-duty vehicles increased for the M25 site but dropped at the M6 site during the crisis, yet both sites saw a decrease in average speeds. The changes in vehicle composition are discussed below.

On the 7th September the flows reached higher levels than they did on the 14th September. On the 7th September maximum flows reached about

2100-2400vph for the offside-1 lane and 2700-3000vph for the offside lane. On the 14th September maximum flows only reached about 1800-2100vph.

Vehicle Composition

Figure 6.11 to Figure 6.14 compare the vehicle composition for the M25 and M6 before and during the crisis. Only one carriageway at each site is plotted. The vehicles are categorised by length. There are four length bins, which are:

- Category 1 – vehicles less than 5.2m (cars, taxis);
- Category 2 – vehicles less than 6.6m and greater than or equal to 5.2m (minibus, light goods vehicles);
- Category 3 – vehicles less than 11.6m and greater than or equal to 6.6m (medium/heavy duty vehicles, etc.); and
- Category 4 – vehicles greater than or equal to 11.6m (heavy-duty vehicles, articulated vehicles).

For the M25 (Figure 6.11 and Figure 6.12), the composition of vehicles is similar for both directions. The drop in total flow was about 65-66 per cent, but the proportion of vehicles in length category 1 went from 58 per cent of the total vehicles before the crisis to 78 per cent during it. This suggests that, although the total flow dropped, it was heavy-duty vehicles that represented the greater drop, rather than car traffic.

For the M6 (Figure 6.13 and Figure 6.14) again the composition of vehicles was similar for both directions. The drop in total flow was about 39-40 per cent. For this motorway, it was the proportion of vehicles in length category 4 that rose, from about 14 per cent before to 20 per cent during the crisis. Length category 1 droped from about 77 per cent before to about 68 per cent during the crisis. This suggests that, although the flow dropped for all the length categories, it was car traffic that represented the greatest drop.

For the M27 the drop in flow was 38 per cent. However, the vehicle composition hardly changed throughout the study period (not shown).

It is a matter for speculation as to why the vehicle compositions changed differently for each of the three sites. For example, it may be that there was an UK wide reduction in long-distance car journeys relative to goods vehicle movements. This would account for the greater reduction in cars relative to heavy-duty vehicles on the M6 near Birmingham. It may be that the M25 and M27 does not attract such a high proportion of long distance traffic.

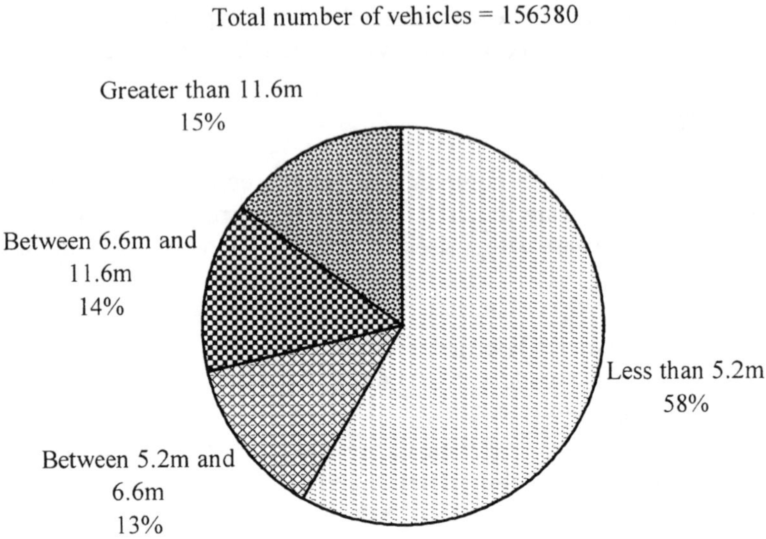

Figure 6.11 Vehicle composition of the M25 anti-clockwise carriageway between junctions 11 and 10 on 07/09/00

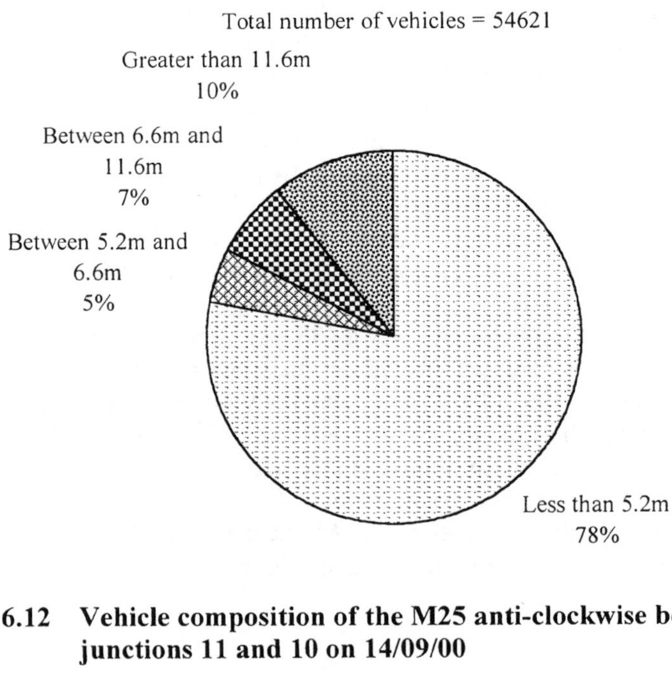

Figure 6.12 Vehicle composition of the M25 anti-clockwise between junctions 11 and 10 on 14/09/00

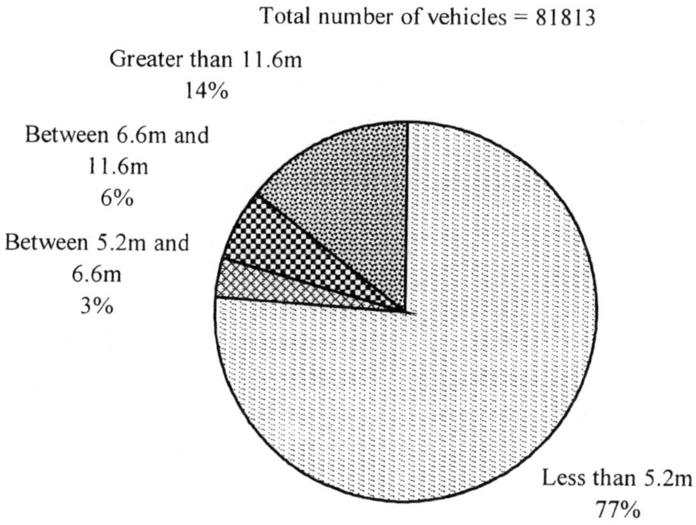

Figure 6.13 Vehicle composition of the M6 northbound carriageway between junctions 7 and 8 on 07/09/00

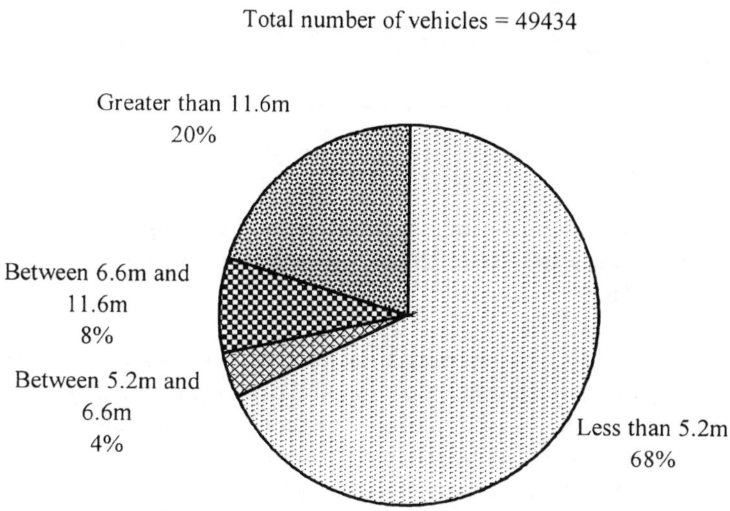

Figure 6.14 Vehicle composition of the M6 northbound between junctions 7 and 8 on 14/09/00

Estimated Journey Times

Journey times can be estimated from the MIDAS speed output. This is the one-minute average speed of all vehicles passing over a loop. The exact locations of the loops are known and they are spaced approximately 500m apart. By extrapolating the speed from each loop location over the space between loops (i.e. +/- 250m), an algorithm can be used to estimate a journey time. Due to the averaging process in MIDAS, the journey time algorithm can underestimate journey times in congested conditions.

The journey times are calculated between Junctions 10 and 16 in the clockwise direction on the M25 for the 7th and 14th of September 2000. They have also been calculated for the 21st September 2000, so that comparison can be made of the after effects of the fuel crisis.

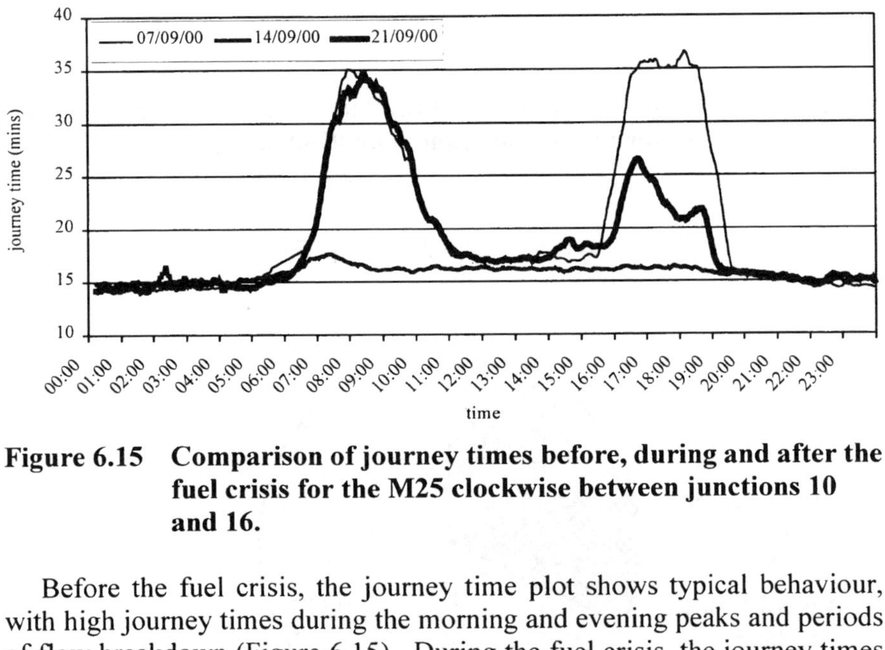

Figure 6.15 Comparison of journey times before, during and after the fuel crisis for the M25 clockwise between junctions 10 and 16.

Before the fuel crisis, the journey time plot shows typical behaviour, with high journey times during the morning and evening peaks and periods of flow breakdown (Figure 6.15). During the fuel crisis, the journey times remained relatively constant, at between 15 and 18 minutes, throughout the day. This indicates that there was no flow breakdown for that day. Journey times rapidly returned to normal, with the exception of the evening peak period on the 7th September, where journey times were significantly longer than those after the fuel crisis. The curve of journey time on the 21st September is more typical of journey times throughout the day (i.e. a sharper peak in the morning peak period, with a longer, lower peak in the evening peak period). Journey times are very sensitive to variations in traffic conditions, so the large journey time experienced on the 7th September could be due to a minor, unreported incident or a slow moving vehicle.

Exhaust Emissions

The exhaust emissions produced by a stream of traffic depend on the volume of the traffic, the types of vehicle present, and the emission rates of each type of vehicle. Vehicle emission rates depend to a large extent on the way in which the vehicles are operated. Currently in emissions modelling, the most widely used operational parameter to define the emission rate is average vehicle speed, and the characteristic variation of emissions with average trip speed is well known. Emissions modelling was conducted on only one of the three sites, the M25 site 4767, between Junctions 10 and 11.

In this study, the emissions from 16 categories of vehicle were estimated using the up-to-date speed-dependent empirical emission functions (based on polynomial regression) which were developed in the European Commission's 4th Framework MEET project (European Commission, 1999). The 16 vehicle categories are listed in Table 6.1. Loop data available from the MIDAS system allows vehicles to be categorised by length. The length categories have been mapped on to the MEET categories and are shown in Table 6.1.

Table 6.1 Vehicle categories represented in the emission calculation

Vehicle category	Vehicle type	Fuel	Weight (t)	Technology level	Engine size	MIDAS length category
1	Passenger cars	Petrol	<2.5	ECE 15.04	<1.4	1 (<5.2m)
2					1.4-2.0	
3					>2.0	
4				Euro 1	<1.4	
5					1.4-2.0	
6					>2.0	
7		Diesel		Uncontrolled	<2.0	
8					>2.0	
9				Euro 1	All	
10	Light goods vehicles	Petrol	<3.5	Uncontrolled	All	2 (5.2-6.6m)
11				Euro 1		
12		Diesel		Uncontrolled		
13				Euro 1		
14	Heavy goods vehicles	Diesel	3.5-16	All	All	3 (6.6-11.6m)
15			16-32			4 (>11.6m)
16			32-40			

The MEET model accounts for the vehicle-to-vehicle variability of emissions by relying on data obtained using a large number of vehicles and a variety of driving cycles. The model was used to calculate emissions per vehicle kilometre of carbon monoxide (CO), volatile organic compounds (VOC), oxides of nitrogen (NO_x), carbon dioxide (CO_2), and total particulate matter (PM) from Diesel engines. Consideration was only given to 'hot' exhaust emissions (i.e. those emissions produced when an engine is at its normal operational temperature).

The minute-by-minute data from MIDAS were processed to provide hourly data. For each hour of the study period, the emission rate of a given pollutant from the traffic travelling in both directions on the motorway section was determined using the following equation:

$$E_h\ (g/km) = \sum_{l=1}^{l=8} \sum_{v=1}^{v=16} e_{l,v,s} \cdot x_{l,v} \cdot f_l$$

Where: E_h is the hourly emission from the traffic on the link

l is the lane number (junction 10-11, lanes 1-4; junction 11-10, lanes 1-4)

$e_{l,v,s}$ is the emission (g/km) from vehicle category v, in lane l, travelling at traffic speed s

$x_{l,v}$ is the proportion of traffic in lane l which is in vehicle category v

f_l is the traffic flow in lane l

Daily total emissions were calculated by summation of the appropriate hourly totals.

There were missing data from the MIDAS system for periods during the 2nd, 6th, 17th and 22nd September. Because daily total emissions could not be calculated for these dates, they were excluded from the analysis.

MIDAS only records the length of vehicles, aggregated across all lanes. Thus, some assumptions and estimations are required to allow the modelling of emissions. Table 6.1 shows the assumed mapping between vehicle length and MEET category.

From an earlier study conducted by TRL Limited, details of the traffic composition in each lane of the M25 were available from an Individual Vehicle Data (IVD) hourly record for a location between junctions 13 and 14 during the period 3rd to 30th April 2000. Thus, the IVD record was used to provide the best estimate of traffic composition between junctions 10 and 11 of the M25 during the month of September. Typical diurnal patterns of traffic composition were obtained separately for weekdays,

Saturdays, and Sundays. However, the MIDAS record of traffic composition for the section of the M25 between junctions 10 and 11 clearly indicates that there was an increase in the proportion of passenger cars in the traffic flow (Figure 6.11 and Figure 6.12).

The vehicles recorded by the IVD are also divided in the same four length categories provided by MIDAS and the mapping shown in Table 6.1 were used.

Because the MIDAS system of vehicle categorisation is rather general, a number of additional assumptions were required to provide the more detailed breakdown of traffic composition required for the emission estimate. The split between petrol and diesel passenger cars, and the proportions of petrol cars in each technology level and engine size band, were determined from national vehicle registration statistics (SMMT, 1999). Diesel cars were categorised by technology level and engine size using information presented in the MEET report (European Commission, 1999). The MEET report was also used to classify light goods vehicles by fuel type and technology level. As all fuel types were in short supply during the fuel crisis, it was assumed that the proportions of petrol and diesel vehicles within each MIDAS category did not change.

Although MEET provides equations to calculate emissions for heavy goods vehicles in two sub-16t bands (3.5t-7.5t, and 7.5t-16t), no information was available to enable the proportion of vehicles in these two bands to be estimated. Consequently, the emission function for vehicles in the 7.6t-16t band was applied to all sub-16t vehicles. Vehicles weighing more than 16t were split into two weight bands (16t-32t and 32t-40t) using different national transport statistics (DETR, 2000). It was assumed that all petrol passenger cars and light goods vehicles registered after January 1st 1993 were fitted with a catalyst and therefore in the Euro 1 category.

Total Daily Emissions

The total daily emissions of CO, VOC, NO_x, CO_2, and Diesel PM during the study period are shown in Figure 6.16 to Figure 6.20. In each figure, the total daily traffic flow (for all lanes, in both directions) is also shown. It can be seen that weekday emissions of each pollutant began to drop markedly after 11th September, and did not return to their original level until around 20th September.

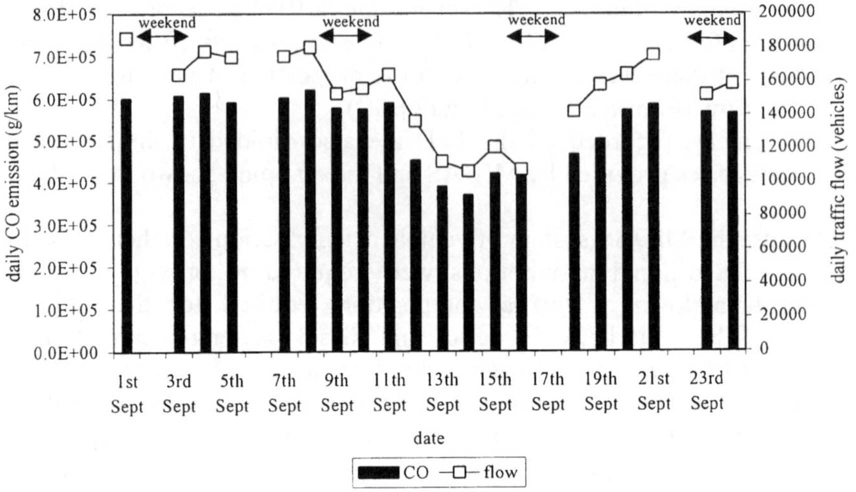

Figure 6.16 Total daily CO emissions and traffic flow during the study period for the site between junctions 10 and 11 of the M25

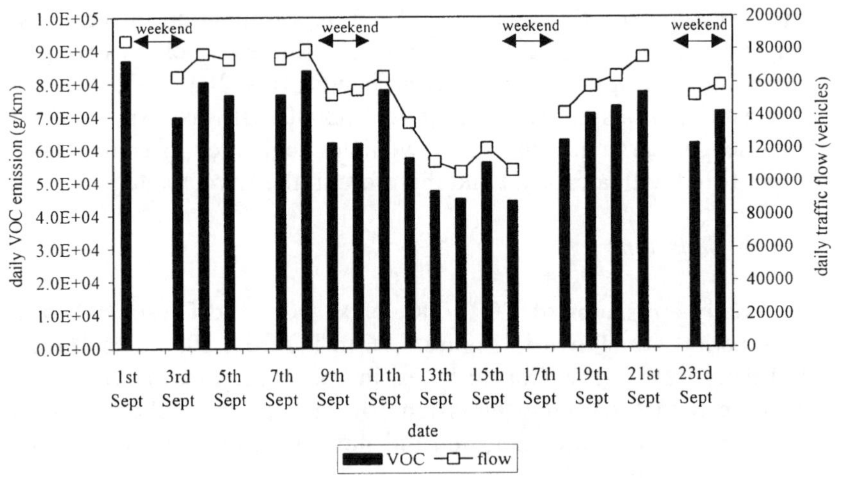

Figure 6.17 Total daily Volatile Organic Compounds emissions and traffic flow during the study period for the site between junctions 10 and 11 of the M25

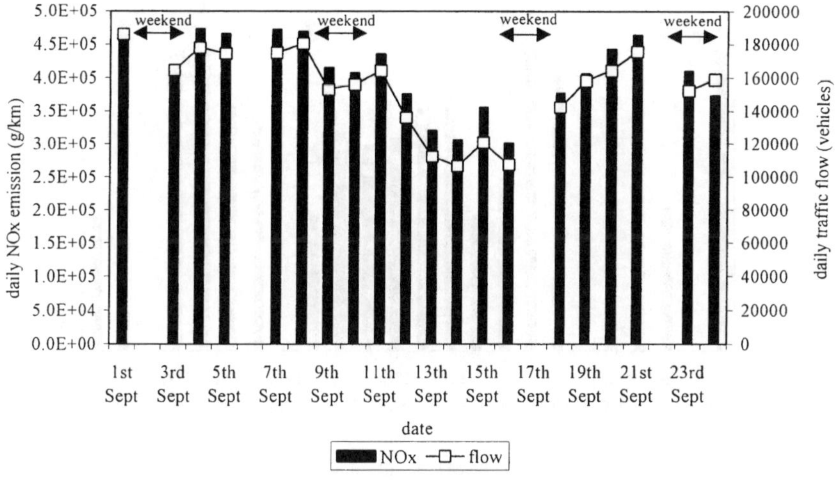

Figure 6.18 Total daily NO_x emissions and traffic flow during the study period for the site between junctions 10 and 11 of the M25

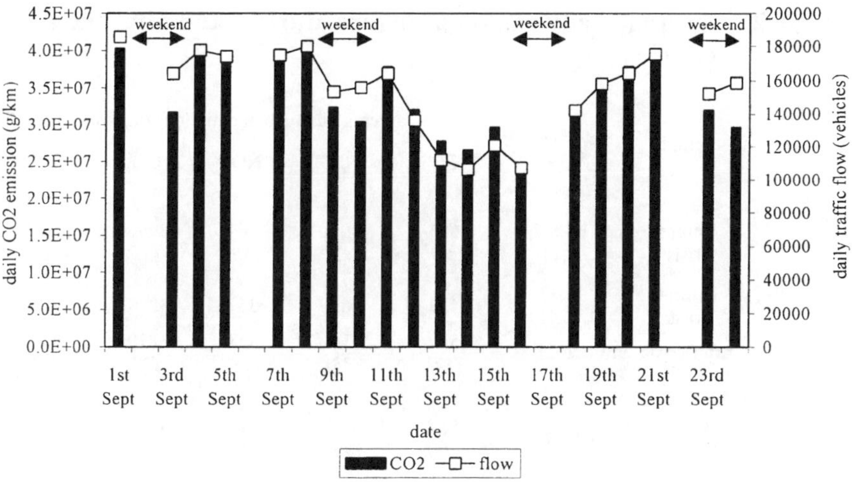

Figure 6.19 Total daily CO_2 emissions and traffic flow during the study period for the site between junctions 10 and 11 of the M25

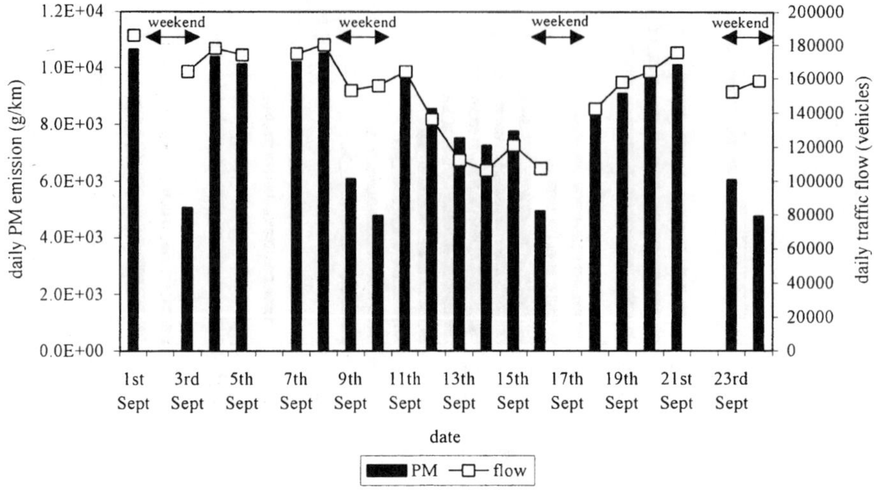

Figure 6.20 Total daily diesel Particulate Matter (PM) emissions and traffic flow during the study period for the site between junctions 10 and 11 of the M25

Table 6.2 Average daily emissions before, during, and after the fuel crisis

Day type	Period	Dates (Sept)	Average daily total emissions (kg/km) CO	VOC	NO_x	CO_2	Diesel PM
Weekday	Before crisis	1,4,5,7,8,11	603 *(39)**	80.4 *(42)*	464 *(30)*	39900 *(31)*	10.3 *(27)*
	During crisis	12,13,14,15,18,19	433	56.5	357	30500	8.11
	After crisis	20,21	579 *(34)*	75.1 *(33)*	453 *(27)*	38700 *(27)*	9.89 *(22)*
Weekend	Before crisis	3,9,10	587 *(40)*	64.5 *(46)*	411 *(37)*	31600 *(31)*	5.32 *(8)*
	During crisis	16	420	44.2	301	24100	4.94
	After crisis	23,24	565 *(35)*	66.5 *(50)*	392 *(30)*	30800 *(28)*	5.40 *(9)*

* The number in brackets indicates the percentage by which the emission level during the relevant period was higher than that during the fuel crisis.

The average daily emissions immediately before, during, and immediately after the fuel crisis are shown in Table 6.2. The extent to which emissions of gaseous pollutants on days before and after the crisis were higher than emissions on the days during the crisis was dependent upon the type of day and pollutant, but generally varied between around 30 per cent and 50 per cent. However, emissions of PM from diesel vehicles were 20-30 per cent higher on weekdays, and around 10 per cent higher on the weekends.

Emissions Due to Changes in Speeds

In Figure 6.16 to Figure 6.20 the decrease in emissions corresponds to a decrease in traffic flow for each pollutant. This suggests that the effect of the crisis on emissions was primarily governed by the resulting changes in flow, with the changes in traffic speed playing a minor role.

Whilst total hourly CO emissions correlated well with traffic flow in free-flow conditions, the importance of traffic speed in determining emissions becomes much more pronounced in more restricted traffic conditions. At flows greater than around 800 vph, flow breakdown occurs more frequently, and the resulting reductions in speed tends to have a dramatic effect on emissions. However, the effect of flow breakdown on emissions is dependent on the free-flow traffic speed. On the M25 for this period of study, the free-flow traffic speed at 800 vph was around 85 km/h for lane 1. From Figure 6.21 (which is specific to a particular type of vehicle but can be used for illustration), it can be seen that this speed corresponds to a point near to the minimum of the CO emission curve. The reduction in speed in lane 1 associated with higher traffic flows will have therefore tended to produce higher hourly emission totals than might have been expected given the traffic flow. In lanes 3 and 4, the free-flow traffic speed was around 120 km/h, a speed that corresponds to a relatively high CO emission rate. Consequently, the reductions in speed in lanes 3 and 4 at higher traffic flows tended to result in lower hourly emissions than might have been expected given the traffic. Slightly different patterns resulted for each pollutant, due to differences in the speed-emission relationships.

The relative extent to which changes in traffic flow and changes in traffic speed contributed to the reductions in emissions during the crisis is difficult to determine with accuracy. Here, in order to isolate the effects of changes in traffic speed, the daily emission of each pollutant on each day of the study was re-calculated after normalising for traffic flow. The traffic speed data were not changed. Average traffic flows for weekdays and weekends were calculated for the period of the crisis, and the appropriate values were assigned to the days before, during, and after the crisis. As speed and flow are related, this approach is not entirely reliable, but the exercise was viewed as a first approximation, and hence no further adjustments were made to the data.

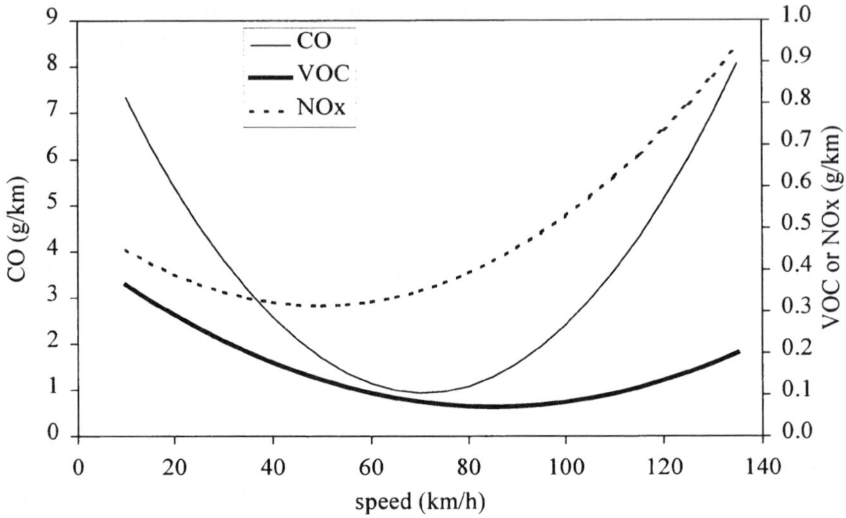

Figure 6.21 CO, VOC, and NO_x emission functions for Euro 1 petrol cars with engine size 1.4-2.0 litres (European Commission, 1999)

Table 6.3 Percentage changes in daily emissions due to the effects of changes in speed

day type	period	% change in emissions due to changes in speed alone				
		CO	VOC	NO_x	CO_2	Diesel PM
weekday	during crisis compared with before crisis	0	-6	+6	+2	0
	during crisis compared with after crisis	+1	-2	+4	+2	+1
weekend	during crisis compared with before crisis	+10	+1	+5	+5	+5
	during crisis compared with after crisis	+10	-4	+8	+6	+6

The results of the calculation are shown in Table 6.3. Clearly, the percentage changes shown here are considerably lower than the overall changes in emissions presented in Table 6.2. This indicates that, as suspected, the change in traffic speed during the crisis did not have a strong bearing on its effect on emissions. From Table 6.3, it appears that, overall, the changes in speed alone during the fuel crisis tended to contribute to a small increase in emissions of CO, NO_x, CO_2, and PM, and a small decrease in emissions of VOCs.

Discussion

During the fuel crisis, the normal rules of supply and demand did not occur (i.e. the price of fuel did not increase dramatically due to demand outstripping supply). Fuel was available primarily on a 'first come – first serve' basis, rather than on a formal rationing system, such as a weekly allowance and special dispensation for public transport and essential services.

Additionally, there was uncertainty in both the availability of fuel and the duration of the crisis. There was also surprise in the speed with which the crisis struck.

The physical effect on the inter-urban road network was two-fold. First, there was a dramatic decrease in traffic volume and secondly there was a decrease in desired speed (for a given flow level). The removal of congestion and reduction in speeds are desirable for many reasons. However, it is unclear how the fuel crisis can assist in providing ideas for achieving these two beneficial effects.

The Removal of Congestion

Congestion was removed by reducing the demand for the inter-urban road network. The demand was reduced by reducing the availability of fuel. The fuel crisis does not inform the debate on the effect of any form of increase in the cost of travel as the cost of travel for road vehicles did not increase. The fuel crisis does not provide any useful data for predicting the degree to which, for example, fuel costs would have to increase to replicate the traffic conditions found during the fuel crisis. Neither can it act as a predictor of the level of road pricing required. Road pricing would, in any case, have a different effect on travel behaviour (e.g. route changing to avoid tolls). Access control would probably have a similar effect to road pricing. However, there may have been a modal shift from single to multiple occupancy vehicles or a shift to modes of transport that did not rely on oil-based fuels but this is beyond the scope of this study.

The decrease in congestion evident during the fuel crisis could only be induced through fuel rationing. Even then, the effects would be different due to the measure being permanent or semi-permanent, rather than the (at the time) unknown duration of the fuel crisis. It would be likely that modal shifts would occur towards multiple occupancy vehicles and public transport (It is assumed that the rationing of fuel to public transport would be less restrictive). However, it is extremely unlikely that fuel rationing would be considered, except in times of national emergency.

Decrease in Desired Speed

The ability to influence drivers so that they decrease their desired speed would be useful for two main reasons. First, it would have a positive impact on safety and the severity of accidents. Secondly, it may also assist in smoothing the flow of traffic and postponing the onset of flow breakdown. However, the decrease in desired speed observed during the fuel crisis was probably due to drivers wishing to conserve fuel. This study shows that the average desired speed of drivers may be reduced, at least in the short-term, by up to 10 kph (6 mph), due to a fuel crisis. This study only allows the prediction of speed reduction due to severe fuel rationing. 10 kph is a small reduction in speed and, whilst desirable, it is unlikely that fuel rationing would be considered as a viable means of reducing average speed. It would not be a useful traffic management tool.

It is also clear that road pricing or access control would not, in itself, cause drivers to reduce their desired speed. It may be possible that an increase in the cost of fuel would cause drivers to attempt to conserve fuel by driving more slowly but the fuel crisis does not provide any information to allow prediction of the effect on desired speed due to the cost of fuel.

Conclusion

Analysis of the chosen sites on the M25, M6 and M27 show that the level of overall flow for the period of the fuel crisis (between the 7th and 15th September 2000) was down. The total daily flow for the M25 on one particular day (14th September 2000) was down by about 66 per cent compared to the same day of the previous week (7th September 2000).

There was no flow breakdown experienced during the period of the fuel crisis on any of the three motorways; thus the speeds were higher during the fuel crisis. As a result of the absence of flow breakdown, journey times were lower.

The three motorways experienced an overall drop in flows, but they all experienced different changes in vehicle composition. For example, the

M25 experienced a drop in numbers of HGVs, whereas the M6 and M27 both showed little decrease in the amount of HGVs.

Analysis of the MIDAS one-minute average speed data in each lane suggests that the speeds at which drivers chose to drive for a given flow level were lower than is usually chosen. This was probably in order to conserve fuel.

Daily emission rates from traffic travelling between junctions 10 and 11 on the M25 dropped markedly during the period of the fuel crisis. For CO, HC, NO_x and CO_2, the decrease varied between around 30 per cent and 50 per cent. Emissions of PM from diesel vehicles were 20-30 per cent lower on weekdays during the crisis, and around 10 per cent lower on the weekend during the crisis.

For each pollutant the effect of the crisis on emissions was primarily governed by the changes in traffic flow. The effects of changes in speed were quite complex but, on balance, comparatively small.

During the week following the fuel crisis, traffic behaviour rapidly returned to normal. No lingering effects were apparent.

It should be noted that this analysis only covers one site from each of three specific motorways and has been based upon the analysis of time averaged MIDAS traffic data.

Due to the uniqueness of this event, and its temporary effect, these results cannot be used to predict the effects of any form of permanent traffic management scheme or changes in the cost or availability of fuel.

Acknowledgement

The authors would like to acknowledge the Highways Agency for allowing TRL Limited to use raw MIDAS data for this work.

References

DETR (2000). *Transport Statistics Great Britain 2000 Edition.* London, TSO.

European Commission (1999). *MEET - Methodology for calculating transport emissions and energy consumption.* Office for Official Publications of the European Communities, L-2985 Luxembourg, 1999.

SMMT (1998). *Motor Industry of Great Britain 1997 World Automotive Statistics.* A statistical publication of the Society of Motor Manufacturers and Traders Limited, London.

7 Travel Behaviour of Car Users During the UK Fuel Crisis and Insights into Car Dependence

KIRON CHATTERJEE AND GLENN LYONS

Introduction

For one week in September 2000 the pumps ran dry in filling stations across the UK. Motorists were unsure when they would be able to refill their vehicles and journeys by car that would previously have been taken for granted were scrutinised and assessed in terms of their necessity and efficiency. A national opinion poll found that 29 per cent of the GB population with a car available to them said they ran so low on petrol they had to stop using their car (ICMa, 2000).

Cars are a central part of most people's lifestyles in the UK and it is not surprising that the public react with caution to transport policies that could restrict use of their cars. Opportunities to explore the response in practice of car users when use of their cars is compromised are rare. The fuel crisis presented such an opportunity on an unprecedented scale.

This chapter describes the results of a mailback questionnaire survey of car users undertaken immediately after the fuel crisis. It starts by briefly summarising what is known about car user behaviour and what further understanding might be gained from a survey at the time of the fuel crisis. There are then some details of the design and distribution of the survey. Chapter 11 (Lyons and Beecroft, 2002) has a detailed account of how the survey was conducted. The chapter then explains how the survey information was processed into a form amenable to analysis and describes the sample of responses received. The main part of the chapter describes the results of the survey. The interpretation and implications of the results are discussed before the chapter concludes with recommendations on policies to reduce car use.

Car Dependence

Increasing car use leads to congestion, delays, pollution, accidents, higher costs as well as reduced opportunities for those without access to a car. The need to develop more sustainable patterns of mobility, whilst not compromising economic prosperity and the choices available to individuals, continues to be a goal for policy-makers both in the UK and many other countries. The UK government's ten year plan for Transport (DETR, 2000) aims to increase the share of travel undertaken by alternatives to the car, whilst recognising that there will continue to be a growth in travel demand and car ownership.

A report on car dependence noted that:

> At the time of first purchase, the car may be seen as a luxury. However, once bought, it encourages changes in behaviour and circumstances which in effect turn it into a necessity. Car dependence grows, rather than simply existing. (Goodwin, 1995)

Nevertheless, it is considered that about 20 per cent of car trips are marginal and could easily be undertaken by other means. The report suggests that:

> In order to target policies effectively, further research will be needed to identify the characteristics of the trips in ... more detail.

Examining the behaviour of car users during the fuel crisis should be able to shed light on the car trips that are most amenable to change and what might be done in policy terms to bring about sustained change to these trips. It should also deepen our understanding of the constraints and difficulties of bringing about change to the other 80 per cent of trips made by car.

It is increasingly recognised by those seeking to understand and predict travel behaviour that travel needs to be viewed in the context of the activities that people pursue. Bhat and Koppelman (2000) have found that studies of travel behaviour taking an activity-centred approach have given an insight into people's routines, household interactions and constraints, and the effects of these on travel. Such studies have usually focussed on specific groups of the population or specific circumstances, and have rarely been able to examine what happens when a major intervention occurs. The fuel crisis provided the opportunity to better understand the ways car users prioritise their activities/trips and reorganise them when they are unable to follow their intended routine.

Car User Survey

Chapter 11 explains the choice of survey method and provides a detailed account of how the survey was conducted. In short, there was a need to capture details of changes to behaviour that occurred during the fuel crisis very quickly before memories faded. Given the absence of any advance warning of the fuel crisis (and hence the lack of any survey planning) a method was required that would yield a large number of responses in a short space of time without the use of a trained survey team. A self-completion questionnaire was used for the survey as it met these criteria as well as enabling detailed accounts of behaviour to be recorded by respondents.

The fuel crisis affected all forms of transport (for example, public transport experienced overcrowding and service cancellations due to drivers not being able to get to work). The aim of the survey, however, was to find out how car users changed their travel behaviour as a result of the fuel shortage. Car use involves those who drive and those who are passengers and includes people who have access to a car at home and those who use pool cars, hire cars or get lifts. The questionnaire was designed to obtain details about all these types of car use. (Private van use was considered in the same way as car use.)

The questionnaire sought to find out about car users' travel behaviour by asking respondents to record their usual travel behaviour and how this had been affected by the fuel crisis. An alternative approach would have been to design the questionnaire as a travel diary. Travel diaries involve subjects recording a chronological account of individual trips and activities. To have been appropriate in this situation, a 'before' survey would need to have been conducted to capture usual travel behaviour. Furthermore, travel diaries would need to have been distributed before the start of the fuel crisis so they could be completed during it. Our approach of asking about change to usual travel behaviour was believed to be appropriate given that the questionnaires were distributed a few days after the fuel crisis. We expected that people would still be able to recollect changes made.

As well as asking for details of changes made to travel, respondents were also asked to give details of any effects the changes had on other activities and household routines. The aim was to get an understanding of the constraints on making changes and the implications of making changes on other aspects of people's lives.

The questionnaire is included at the end of Chapter 11. The main part of the questionnaire centred on four travel purposes: the journey to/from work (commuting); journeys made whilst at work (business); escorting children to/from school or pre-school (school); and grocery shopping (groceries). Table 7.1, which is based on national transport statistics (DETR, 2001), shows that travel for these purposes comprises a substantial proportion of all travel undertaken. Commuting, business, escort education and shopping account for 46 per cent of all car trips and 45 per cent of car travel distance.

Table 7.1 Average annual travel by purpose in GB (DETR, 2001)

Purpose	Travel by all modes		Travel by car	
	% of trips[a]	% of distance[b]	% of trips[c]	% of distance[d]
Commuting	16	20	18	19
Business	4	10	4	11
Education	6	3	3	2
Escort education	5	1	4	2
Shopping	21	13	20	13
Leisure	31	40	29	40
Other personal and escort	18	13	21	14

a Average annual total trips by all modes = 1,046
b Average annual total distance by all modes = 6,806 miles
c Average annual total trips by car = 646
d Average annual total distance by car = 5,546 miles

The questionnaire had eight sections.

- *Section 1*: asked for details of the respondent and their household, including age, gender, occupation, available cars and whether cars refuelled at start of fuel crisis week.
- *Sections 2 to 5*: applied to the four travel purposes mentioned above and asked for details of usual routine, whether any change made to usual routine (including details of change made and reasons why not if no change made), whether change had been made before in last six months and whether consideration would be given to making change again in future.
- *Section 6*: asked if other changes had been made to travel and usual routine, including any combining of trips and activities, any cancellations and postponements of trips and activities, any extra use of the phone or Internet and any seeking of public transport information.
- *Section 7*: asked for views on eleven specific statements related to the fuel crisis (e.g. 'did the fuel crisis make you realise you use your car more than necessary').
- *Section 8*: provided space for respondents to note any other things that they wished to say about the fuel crisis and future levels of car use, fuel tax or related issues.

A key feature of the questionnaire is that open response questions were used where respondents were invited to describe in their own words what changes they had made and why. The alternative was to have used closed response questions where respondents were asked to note which of a set of option answers applied to them. Without time to pilot the questionnaire it was not possible to identify the range of different changes to car use that were made during the fuel crisis and to establish appropriate answer options. Closed response questions would have proved vastly more manageable and straightforward in analysis terms but would have compromised the aim of capturing the full breadth of circumstance and types of change that were made.

Mailback questionnaires were distributed to the public door-to-door, in town and city centres and via petrol filling stations, primary schools and employers. They were distributed in five areas of England (Hampshire/Wiltshire, West Yorkshire, London Borough of Hillingdon,

Leicester and Hertford). A total of over 10,000 mailback questionnaires were distributed. In addition to the distribution of mailback questionnaires, a duplicate version of the questionnaire was created for the web. This allowed individuals with access to the Internet from any physical location in the UK to participate in the survey. This chapter only presents results of the mailback questionnaire.

Chapter 11 has a more detailed account of the survey distribution and an analysis of the response rates achieved for different distribution methods.

Fuel Crisis Survey Sample

Overall Sample

A total of 1,659 usable responses was received. A cut off date for usable responses of 19 October (34 days after the end of the fuel crisis on 15 September) was applied. This was an attempt to minimise the inclusion of responses that would contain distorted or vague recollections of events, behaviour and attitudes during the crisis. In fact, 88 per cent of the 1,659 responses were received within 14 days of 15 September.

Data Processing

A major consequence of having many open response questions was the substantial time required to firstly transcribe the questionnaire responses and secondly process the responses and categorise them to enable quantitative analysis to be carried out. Table 7.2 is a listing of variables created from open response questions.

In categorising the response information in sections 2-6 a compromise had to be made between capturing the enormous variety of behavioural change that occurred and obtaining meaningful insights into the most common changes. To categorise the types of changes made to car use a master-list of 75 different codes was used. The master-list is included at the end of the chapter. The same master-list applied to each of the four travel purposes. The example below for the journey to/from work for respondent #133 illustrates the task that was involved:

> Instead of taking my children to school they cycled there and back. Normally I would take them in the morning and they would bus home in the afternoon. This made it easier for me but my daughter who has a slight

physical disability found cycling hard and aggravated her medical problems. (#133)

This response was coded as type '64' – went directly to work instead of trip chaining. The detailed codes were also grouped into 13 higher level codes (which applied to the four travel purposes and to other travel – these are also shown on the master-list). In terms of the higher level code, #133 was coded as a type '3' trip chaining change.

Sample Representativeness

The characteristics of the sample were examined carefully to see how representative it was of the general car user population. It was not expected that the sample would be fully representative for a number of reasons. Random sampling of the car user population was not undertaken. Although different areas of the country and different places within each area were targeted the distribution will not reflect the national distribution. Of those receiving a questionnaire, certain types of individual will have been more likely to return them. There is also the likelihood that those with strong opinions on the subject of the survey will have been more likely to respond.

Table 7.3 presents characteristics of the fuel crisis sample alongside comparative figures for the general population. The figures for the general population are drawn from a variety of different sources and in some cases apply to all of the population, in other cases to car drivers and in one case to all of the population in employment.

The fuel crisis sample had a higher proportion of females and people aged between 35 and 54 than the general population of car drivers (differences are statistically significant at the 99.9 per cent level). It had more people with occupation types 1-4 than the general population in employment and fewer people with occupation types 5-9 (difference in occupation type distributions is statistically significant at the 99.9 per cent level).

Table 7.2 Variables created from open response questions

Variable	Definition	Categories
Section 1		
Occupation major group	Classification of occupation according to standard occupational classification 2000 major groups. See ONS (2000).	0=unknown, 1=manager,... 13=retired
Respondent status	Respondent's status in household (all house members considered)	0=unknown, 1=on own,... 9=with other adult co-residents
Household type	Life-cycle based categories as used by Jones et al. (1983)	0=unknown, 1=young adult(s) w/out children,... 8=retired persons
Sections 2-5		
Change made	Changed travel on any day because of fuel crisis?	0=n/a or unknown, 1=yes, 2=no
Reason not changed	Reason given for not making any change to travel for this purpose	0=n/a or unknown, 1=on holiday,... 21=conserved fuel elsewhere so that car could be used
Type of change made*	Type of change made on occasion when usual routine not followed	0=n/a or unknown, 1=bus,... 84=travel as normal but destination activity not possible due to fuel crisis
Frequency change made*	Number of days when change made	0=n/a or unknown, 1=on one day, 2=on at least one day
Reason for making/not making change again*	Main reason for making or not making change again in future	0= not applicable/unknown, 1=cost,... 44=better balanced lifestyle

Table 7.2 Variables created from open response questions

Variable	Definition	Categories
Section 6		
Other travel purpose*	Purpose of other type of travel made on a regular basis	0=not applicable/unknown, 1=visit family/friends,… 25=go to concert/theatre
Other travel impact*	Impact of fuel crisis on other type of travel	0=n/a or unknown, 1=cancelled,… 24=linked to other activity
Reason used phone more	Reason made greater use of phone/Internet during fuel crisis	0=n/a or unknown, 1=find out where fuel available,… 23=rearranging business matters

* Allowance made for details of more than one change to be considered for each respondent for each travel purpose.

Table 7.3 Comparison of fuel crisis sample with general population

Characteristic	Fuel crisis sample (%)	General pop. (%)
Gender		Lex[a]
Male	50	60
Female	50	40
Age group		Lex[a]
17-34	23	30
35-54	53	43
>54	24	28
Standard occupational classification of respondents in employment		LFS[b]
Manager (1)	15	14
Professional (2)	30	12
Associate (3)	17	13
Administrative (4)	21	13
Skilled trades (5)	5	12
Personal service (6)	4	7
Sales/customer (7)	3	8
Operatives (8)	3	9
Elementary (9)	3	12
'Which of the following best describes where you live?'		OPCS[c]
Remote	5	
Village	34	
Town	33	
City suburb	21	
City central	7	
Remote, mainly rural		11
Mixed urban/rural and accessible rural		20
Towns		25
Metro		44

Table 7.3 Comparison of fuel crisis sample with general population

Characteristic	Fuel crisis sample (%)	General pop. (%)
Respondent's car use		NTS[d]
Car driver	96	
Household car but not regular driver	1	
No household car	3	28
No. of cars in household for respondents in household with a car		NTS[d]
1	43	62
2	45	31
>2	12	7
Annual mileage of respondent		Lex[a]
Average	9600	9300
'I support government in not reducing fuel taxes as a result of the crisis'		ICM[e]
Agree or strongly agree	34	40
Disagree or strongly disagree	57	55

a Lex figures apply to a 1998 GB survey of 1,297 randomly selected car drivers (defined as driving at least once a month) (Lex, 1999).
b LFS figures apply to UK population in employment in Spring 2001 (LFS, 2001).
c OPCS figures are percentage of population of England and Wales in 1991 living in different district types (OPCS, 1992).
d NTS figures apply to GB population in 1999 (DETR, 2001: Table 1.2).
e ICM figures apply to a petrol crisis poll of 514 randomly selected adults in GB carried out on 12/09/00. The question asked was 'do you think the government should reduce the tax on petrol SPECIFICALLY in response to the protest or not?' 55% said yes, 40% said no (ICM, 2000b).

The difference in occupation type distribution was anticipated to some extent as the fuel crisis survey was aimed at car users who have a higher proportion of people in occupation types 1-4 than the general population in employment. However, the effect was more pronounced than desired and may be due to the distribution approach and differences in propensity to return questionnaires between occupation types.

Of the fuel crisis sample, 39 per cent considered where they lived to be either a 'village' or a 'remote area'. In 1991, 31 per cent of the England and Wales population lived in a remote, mainly rural area, accessible rural area or mixed urban/rural area. The differences in definition in the two cases and the subjectivity inherent in the fuel crisis responses make comparison difficult. At face value, it appears that the fuel crisis sample is over-represented by rural dwellers and under-represented by city dwellers.

Of the fuel crisis survey sample, 96 per cent were regular car drivers. Of the sample 1 per cent lived in households with a car but were not regular drivers of a car. Most of these respondents would be regular car passengers. The remaining 3 per cent of the sample lived in households without a car. Many of these will occasionally be car users and therefore were retained in the sample. In comparison to the fuel crisis survey sample, 28 per cent of the GB population live in households where they do not have a car available to them. This shows that the survey was successful in attracting car users.

Of the fuel crisis survey sample, 57 per cent who lived in a household with a car, had two or more cars while this figure is only 38 per cent for the GB population generally (difference is statistically significant at the 99.9 per cent level). The annual mileage driven by car by the fuel crisis sample respondents was very similar to that by GB motorists generally (difference is not statistically significant at the 95 per cent level).

The proportion of respondents believing that government should reduce fuel taxes as a result of the protest is similar for the fuel crisis survey sample and the general GB population (difference is not statistically significant at the 95 per cent level).

There are large differences between the socio-economic characteristics of the fuel crisis sample and the general population but the amount of car travel undertaken and attitudes to the fuel protests are similar. The main focus of the analysis of the survey data concerned the changes made by car users and the circumstances in which they were made. No attempt has been made to predict aggregate changes made during the fuel crisis or forecast future changes to car use for new transport measures. In this context, the representativeness of the sample is not a substantial concern.

Results of the Fuel Crisis Survey

In this section the results of the survey are described. First comparison is made of the number of changes made for different travel purposes and then the main types of changes are identified and examined in some detail. Reasons given for not changing car use are described before completing the results with some overall responses made to car use.

For what travel purposes were changes made to car use?

In the second column of Table 7.4 it is shown for how many of the respondents each travel purpose is applicable. The lowest figure is for school escort travel where only 36 per cent of the sample said that they undertake this type of travel. Of the sample 80 per cent said they travel to work and 54 per cent said they undertake journeys while at work. In the fourth column of table 7.4 (under 'Car users') it is shown how many of the total sample use the car for each travel purpose. This shows that the car is the main mode for all travel purposes, although for journeys to work and school escort travel many of the respondents use other modes.

Figures on changes to car use are shown both for any change to car use and any *major* change to car use. Major change to car use is taken to include trip reduction, trip origin or destination change, trip chaining, train, bus, cycle, walk, other vehicle and car share. The following changes have not been considered to be major changes, as they are likely to be responses particular to the fuel crisis situation, which would not occur as a consequence of transport policy measures:

- Change of route or driving style to save fuel.
- Change of route to avoid fuel protest demonstrations or petrol filling station queues.
- Rescheduled or additional trips made to stock up on food.

Referring to the master-list of types of change made to car use included at the end of the chapter the types of change not considered are those identified as belonging to groups 10 to 13. Unless stated otherwise, in the results that follow these changes to car use have been excluded from consideration.

Table 7.4 Changes to car use for different travel purposes

Travel purpose	Applicable purpose		Car users (A)		Car users making at least one change of any kind		Car users making at least one *major* change	
	No.	% of total	No.	% of total	No.	% of A	No.	% of A
Commuting	1324	80	1065	64	491	46	450	42
Business	894	54	801	48	326	41	321	40
School	598	36	347	21	132	38	128	37
Groceries	1598	96	1555	94	556	36	481	31
Other*	1659[a]	100	1617[b]	97	910	56	910	56

a 1,659 is total sample size
b Assumes 1,617 respondents living in household with car use the car for other journeys.

About 40 per cent of car users for commuting, business and school escort travel made at least one major change to car use. The figure was lower for grocery shopping (31 per cent) and higher for other travel (56 per cent), which mostly comprises leisure travel.

What types of changes to car use were made?

Table 7.5 sets out the types of changes made to car use for the five different travel purposes. The types of change are mutually exclusive, ie. if two trips are chained then this is considered as trip chaining only and not trip reduction. It shows that 12 to 16 per cent of car users made fewer trips for commuting, business and grocery shopping travel and 51 per cent made fewer trips for other travel. For grocery shopping 14 per cent of car users made a trip origin or destination change. These responses involved using more local shops. In some of these cases a change of mode was also involved (usually from car to walking).

For commuting and school escort travel, substantial numbers of car users changed mode and car shared. Trip chaining, where travel for one purpose is carried out while travelling for another purpose, was only a substantial response for grocery shopping.

Table 7.5 Types of changes made to car use

	Travel purpose									
	Commute		Business		School		Grocery		Other	
No. of car users	1065		801		347		1555		1617	
Type of change	No.	%[a]	No.	%[a]	No.	%[a]	No.	%[a]	No.	%[a]
Trip reduction	126	12	246	16	15	4	216	14	826	51
Trip OD change[b]	26	2	8	1	1	0	210	14	39	2
Trip chaining	12	1	13	2	6	2	75	5	18	1
Train	59	6	19	2	0	0	0	0	14	1
Bus	68	6	11	1	11	3	8	1	28	2
Cycle	40	4	4	0	16	5	12	1	32	2
Walk	30	3	13	2	73	21	24	2	68	4
Car share	123	12	16	2	14	4	8	1	46	3
Other vehicle	31	3	5	1	9	3	11	1	0	-
Total[c]	515	-	335	-	145	-	564	-	1071	-
Driving style[d]	54	5	7	1	5	1	2	0	-	-
Trip increase	0	0	0	0	0	0	50	3	0	0
Trip rescheduling[e]	0	0	0	0	0	0	49	3	0	0
Total[c]	569	-	343	-	150	-	665	-	1071	-

a Percentage of car users (number of car users is noted above).
b Trip origin or destination change.
c Some respondents made more than one type of change for the same travel purpose, therefore no figure is given for total percentage. This also explains why the total number of changes for each travel purpose is greater in Table 7.5 than Table 7.4.
d Changes to driving style not noted for other travel.
e Rescheduling trip for earlier in week.

Care must be taken when comparing responses made between travel purposes. The viability of responses depends on the travel purpose. For most parents and guardians, trip reduction is not a viable choice for school escort travel, unless there is no means to get the children to school. Changing the origin or destination is unlikely to be feasible for commuting and school escort travel. It needs to be recognised that the choice alternatives vary according to travel purpose as well as the circumstances and attitudes of an individual respondent.

Changes to car use made by a substantial number of respondents merit more detailed analysis. The following types of change are identified.

- *Commuting*: trip reduction (in particular working form home), train, bus, cycling, walking and car sharing.
- *Business travel*: trip reduction.
- *School escort travel*: walking.
- *Grocery shopping travel*: trip reduction, trip origin-destination change (in particular walking to local shops and driving to local shops) and trip chaining.
- *Other travel*: trip reduction and walking.

What were the circumstances and experiences of the main changes made for commuting?

Table 7.6 sets out details of the characteristics and experiences of the respondents making the six types of changes to commuting identified above. Only a selection of what appear to be the most interesting characteristics is shown.

In some cases the number of respondents making changes varies slightly from those given in Table 7.5. To avoid double counting in Table 7.5, a change was only counted under one category, even if it involved a mixture of modes. For example, changing from car to use a combination of bus and train was considered only as a change to train. In Table 7.6 the same change is counted under both train and bus, so that attention is paid to all usage of the modes during the fuel crisis.

Table 7.6 indicates the following associations between the changes made and the characteristics of respondents.

- *Working from home*: manager/professional and long commute (average of 27 miles).
- *Train*: male, long commute (average of 23 miles) and West Yorkshire respondent.
- *Bus*: low car availability, short commute (average of 12 miles), West Yorkshire respondent and low on fuel at start of fuel crisis.
- *Cycling*: male, short commute (average of 9 miles) and admit to use car more than necessary.

- *Walking*: low car availability, short commute (average of 5 miles), live in town or city, low on fuel at start of fuel crisis and admit to use car more than necessary.
- *Car sharing*: young respondent and short commute (average of 13 miles).

The associations tell us something about the circumstances in which the different responses were made, although further statistical analysis is required to better understand the relationships between the data. The high proportion of West Yorkshire respondents that used public transport is interesting. This may be due to the West Yorkshire Metropolitan area having a better developed public transport network than the other areas surveyed. Incidentally, the proportion of respondents in Southampton and Leicester (two medium-sized cities) using public transport instead of the car was not higher than average.

Of respondents 64 per cent of those who switched to train or bus mentioned that they sought public transport information during the fuel crisis. Only 19 per cent of all survey respondents said they sought public transport information. About half of respondents who used the train or bus, or cycled or walked had done so previously in the last six months. The proportion was lower for working from home and car sharing. For each of the six types of change to car use more than 40 per cent of respondents said they would consider using the option again in future. The proportion was highest for cycling and walking. Working from home tended to only occur for one day, whilst about half of those changing from car use to using the train, cycling, walking or car sharing made the change for at least two days during the fuel crisis week.

Working from home was not always a practice that fitted well with respondents' occupations:

> Cannot do my job without access to the people in many different locations that are impractical for public transport. (#1074)

Table 7.6 Characteristics and experiences of respondents making change to car commuting

Change	All car users	Work from home	Train	Bus	Cycle	Walk	Car share
No. respondents (A)	1,065	67	59	81	43	31	123
Characteristics							
Mean age	43	43	43	43	43	43	41
Mean cars/household	1.9	1.9	1.7	1.6	1.8	1.5	1.8
Mean commute length in miles	15	27	23	12	9	5	13
		Figures below are per cent of respondents (A)					
Male respondent	50	55	66	49	60	23	45
W. Yorkshire respondent	28	28	47	57	14	29	37
Manager or professional	44	66	53	37	44	39	49
Live in remote area or village	40	42	32	31	35	26	41
At least half tank at start of crisis	84	75	73	63	74	65	74
Admit to use car more than necessary	25	31	27	26	40	45	29
Sought public transport info during fuel crisis	22	43	64	64	21	32	22

Table 7.6 Characteristics and experiences of respondents making change to car commuting

Change	All car users	Work from home	Train	Bus	Cycle	Walk	Car share
No. respondents (A)	1,065	67	59	81	43	31	123
	Figures below are per cent of respondents (A)						
Experiences							
Use car or this option anyway but increased use	-	6	2	10	14	19	3
Changed on 1 day only	-	57	37	38	27	10	30
Changed on 2+ days	-	33	46	38	50	52	45
Used option in last 6 months	-	34	46	38	47	55	29
Willing to use option again in future	-	42	53	42	67	68	47
Reason for not using option again							
Cost	-	0	9	21	2	3	2
Time	-	1	17	18	0	6	4
Tiring	-	0	3	2	11	3	0
Use car at work	-	0	6	8	9	6	2
Loss of income	-	0	0	1	0	0	1
Impact on work	-	12	2	2	5	3	1
Lost flexibility	-	1	8	2	0	3	42
Reason for using option again							
Health/fitness	-	0	0	4	16	10	0
Enjoyable	-	0	3	0	23	13	3

Cost and time were often cited as a reason for not continuing to use public transport in the future.

> Needed to save fuel to ensure enough for appointment in Hull... Takes too long to travel. Uncertain of reliability of railways. Too expensive. (#1522)

> Only if I wasn't rushing to get home or didn't have off-site meetings, etc - or shopping to do! (#1095)

The health benefits and enjoyment of cycling and walking were often mentioned as positive reasons to cycle again in future. The loss of flexibility in car share arrangements was mentioned by nearly half of those who car shared as a problem for future use of this option.

> I enjoyed the company on the journey. Would save fuel money. It would not however be possible every day due to work/personal commitments. (#831)

In processing the survey information it was also noted when a respondent mentioned that they vary their travel option and do not have a single way of getting to work that they use every day. Table 7.6 indicates the percentage of respondents who said they sometimes use the alternative anyway and during the fuel crisis week increased their use of it. For train 2 per cent (one out of the 63 respondents) said they sometimes use the train anyway. For the other travel alternatives the values were higher. For walking it was 19 per cent.

It is important to recognise there is variation in the character of the responses made within the six general categories examined above. Taking the example of train, Table 7.7 is a turnover table showing the precise details of the usual commuting mode and the response made. Only 24 of the 63 changes made were from car driving alone to train with 34 of the 63 changes made from car driving to train combined with another mode. Turnover tables are a useful means of describing changes of mode usage between two periods of time (Chatterjee, 2001). In this case the behaviour of people who usually use the train and other non-car modes has not been surveyed, so a full turnover table is not able to be presented. Thorpe et al. (2002: Table 8.3) present a full turnover table for journeys to work/college/school before and during the fuel crisis.

Table 7.7 Turnover table for respondents using train

Usual / Change	Train	Bus & train	Walk/ cycle & train	Car & train	Total
Car alone	24	8	14	12	58
Car with passenger	0	1	0	1	2
Car alone and train	1	1	0	0	2
Car as passenger	1	0	0	0	1
Total	26	10	14	13	63

What were the circumstances and experiences of the main changes made for business travel?

Trip reduction was the main change to car use that occurred for business travel, accounting for 74 per cent of changes made. It involved cancellations and postponements which are likely only to be acceptable responses to a crisis situation. This is borne out by the fact that only 5 per cent of respondents who made a trip reduction response said they had done the same previously in the last six months and only 17 per cent said they would consider the same option again in future. There was no evidence of particular types of business travel (e.g. meetings, visits to clients, collections and deliveries) being targeted for trip reduction during the fuel crisis.

What were the circumstances and experiences of the main changes made for school escort travel?

Table 7.8 sets out details of the characteristics and experiences of the respondents whose children walked to school instead of using the car. These respondents had shorter journey lengths, although the average was still 1.8 miles. They were less likely to live in a rural area, have older children only, have children at more than one school and be in a car share arrangement. They were more likely to admit to use their car more than necessary (47 per cent compared to 37 per cent for all respondents who use the car for school escort travel).

Table 7.8 Characteristics and experiences of respondents whose children walked to school instead of going by car

Change	All car users	Walk
No. respondents (A)	347	73
Characteristics		
Mean journey length in miles	4.1	1.8
	Figures below are per cent of respondents (A)	
Household with older children only	11	5
Live in remote area or village	39	27
Car share arrangement for school journeys	18	7
Have children at more than one school	31	26
At least half tank at start of crisis	84	78
Admit to use car more than necessary	37	47
Experiences		
Changed on 1 day only	-	16
Changed on 2+ days	-	58
Used option in last 6 months	-	36
Willing to use option again	-	52
Reason for not using option again		
Time	-	11
Tiring	-	8
Safety	-	4
Weather	-	14
Disruption to other activities	-	4
Reason for using option again		
Health/fitness	-	4
Enjoyable	-	5

Of the respondents, 36 per cent whose children walked to school had done so previously in the last six months and 52 per cent said they would consider this option again in future. Most made the change for at least two days during the fuel crisis (58 per cent).

The weather was mentioned as a factor influencing future decisions to walk by 14 per cent:

> When my wife and children are sufficiently organised they walk - except when weather inclement or large quantities of books/musical instruments/games kit etc. (#1091)

Positive aspects were mentioned by some respondents:

> After the initial complaints, my eldest daughter seems to enjoy walking with her friends, but we still have reservations about safety. (#520)

> It is good to encourage the children to be independent in this way. (#618)

What were the circumstances and experiences of the main changes made for groceries travel?

Table 7.9 sets out details of the characteristics and experiences of the respondents who walked to local shops, drove more locally or trip chained for the groceries shopping as a change to their usual use of the car. Trip reduction has not been considered, as it involved postponing grocery shopping and managing with provisions at home which is likely only to be an acceptable response to a crisis situation.

Table 7.9 indicates the following associations between the changes made and the characteristics of respondents.

- *Walking to local shops*: low weekly mileage and short distance for small groceries journeys.
- *Driving to local shops*: low weekly mileage, long distance for large groceries journeys and household with children under 16.
- *Trip chaining*: long distance for small groceries journeys, live in rural area and admit to use car more than necessary.

Table 7.9 Characteristics and experiences of respondents making change to groceries shopping car travel

Change	All car users	Walk to local shops	Drive to local shops	Trip chaining
No. respondents (A)	1555	120	55	75
Characteristics				
Mean weekly car miles	190	160	160	190
Mean journey length in miles for large groceries	5.2	5.3	8.7	5.2
Mean journey length in miles for small groceries	2.4	1.5	2.7	3.0
	Figures below are per cent of respondents (A)			
Live in remote area or village	40	34	44	48
Household with children under 16	37	38	45	43
At least half tank at start of crisis	84	76	78	81
Admit to use car more than necessary	26	28	25	37
Experiences				
Changed on 1 day only	-	12	20	59
Changed on 2+ days	-	39	20	11
Used option in last 6 months	-	10	15	19
Willing to use option again	-	22	25	27
Reason for not using option again				
Cost		17	16	3
Time	-	11	5	4
Bulky items	-	8	2	1
Lack of choice	-	9	20	4

Fewer than 20 per cent of the respondents making each of these changes to car use had done so previously in the last six months and fewer than 30 per cent said they would consider using the option again in future. Of those who walked to local shops 39 per cent did so at least twice.

Expense, time taken, difficulty carrying items and lack of choice were regularly mentioned as reasons for not wishing to use local shops again in the future:

> Not so much choice at local for all of our groceries plus things are more expensive. I do support local shops for basic items however. (#6)

> The bags were heavier by the time I reached home than when I bought them. The car is the answer. (#951)

Positive aspects were mentioned by a few respondents:

> Shopping in this way would not be a great inconvenience as reasonable shops - choice and prices - locally within 1 mile. (#559)

What were the circumstances and experiences of the main changes made for other types of travel?

The survey asked respondents to give details of changes made to other travel planned (prior to the crisis) to be made by car during the fuel crisis. As with business travel, trip reduction was the main change, accounting for 83 per cent of changes made. There were 1,442 mentions of trips not being made which equates to nearly one mention for each respondent. Trip reduction involved cancellations (77 per cent) postponements (12 per cent), phoning instead (5 per cent) and making fewer journeys (5 per cent). In most of these instances not making the journey is likely only to be an acceptable response during a crisis situation. There were also 81 mentions of walking instead of using the car.

Respondents were asked to describe the purpose of other travel affected by the fuel crisis. The main purposes mentioned were social/entertainment (46 per cent), visits to friends (38 per cent), shopping other than groceries (8 per cent) and holidays/day trips (4 per cent). Examination was made to see whether the responses varied according to travel purpose. Visits to the cinema were more likely to be cancelled than other travel purposes. Visits to friends were more likely to be postponed, substituted by a phone call or made on fewer occasions than other travel purposes.

What reasons were given for not changing car use?

Table 7.10 identifies the reasons given for not making a change to car use. Reasons were noted (where mentioned by respondent) for all respondents not making any change to car use. The most common reason for not making a change was having sufficient fuel. This is an expected result given that 71 per cent of the GB population with a car available to them said they did not run so low on petrol they had to stop using their car (ICMa, 2000). It is interesting to note that 13 per cent of commuting car users made no change because they were on leave and 31 per cent of business travel car users made no change because they had no journey to make. Only small numbers mentioned that the inadequacy of other modes was the reason for not making any change.

Table 7.10 Primary reason given for not changing car use

Travel purpose	Commuting	Business	School	Groceries
No. respondents not making any change (A)	574	475	215	999
Reason	Figures below are per cent of respondents (A)			
Away from work/school	13	7	2	3
Enough fuel	51	20	47	33
No journey needed	1	31	0	14
Linked to other trip	1	0	7	7
Other modes unacceptable	7	2	5	1
Other reason	7	3	2	3

How many changes did respondents make to car use?

Figure 7.1 shows for how many of the five travel purposes (commuting, business, school, groceries and other) a major change to car use was made. Only a quarter of the respondents made no change and 42 per cent made at least two changes. The number of travel purposes that were applicable to a respondent varied from zero to five. Only 144 respondents noted that they normally use the car for all five travel purposes.

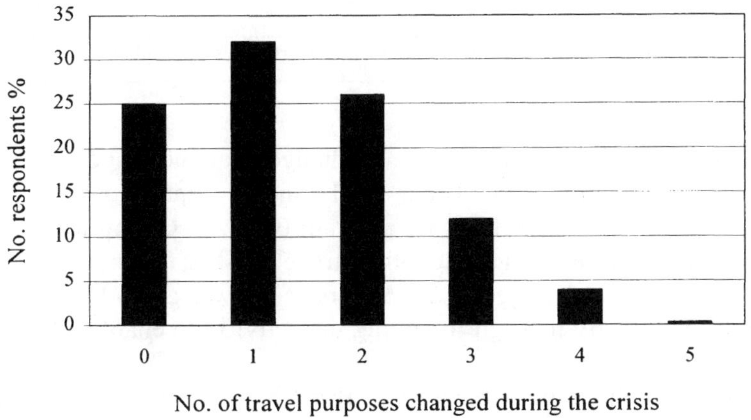

Figure 7.1 Number of travel purposes changed during crisis

How did changes to car use depend on combinations of travel purposes?

Those respondents who use the car for more types of travel might be expected to make changes to a greater proportion of them to conserve fuel. Table 7.11 looks at the most common four combinations of applicable travel purposes (which account for 76 per cent of the survey respondents). The expected numbers of changes in Table 7.11 are calculated based on the average percentage of respondents making a major change to each travel purpose in Table 7.4. Table 7.11 suggests that respondents using the car for four or five travel purposes made about the same proportion of changes as those using the car for two or three purposes.

Table 7.11 Number of purposes of travel changed

Applicable travel purposes	No. of respondents	Expected number of changes	Actual number of changes
G, O	329	286	276
C, G, O	291	375	377
C, B, G, O	490	828	856
C, B, S, G, O	144	297	303

C = commuting, B = business, S = school escort, G = groceries, O = other

Examination of the travel purposes changed for the respondents with the four combinations in Table 7.11 shows that those with the combinations of five (C, B, S, G, O) and four (C, B, G, O) applicable travel purposes made more changes to other travel. Most other travel is for leisure purposes which suggests that leisure travel was given lower priority and modified (mostly through foregoing the activity) to enable other travel to take place.

The four combinations above have also been analysed to see whether within them there are greater than expected numbers of 'flexible' car users (who made many changes) and 'inflexible' car users (who made few or no changes) than would be expected if every respondent had an equal probability of making a change to each travel purpose. The expected numbers of changes were again calculated based on the average percentage of respondents making a major change to each travel purpose in Table 7.4. The analysis, indeed, showed that there are 'flexible' and 'inflexible' car users. There are many possible reasons for this, including the likelihood that some respondents had more amenable circumstances than others to make changes (for example, more alternative modes and local services) and some respondents had more available vehicles and fuel. There will also have been differences in attitudes towards the need to make any change to car use.

An understanding has been gained of the circumstances and experiences surrounding the main types of changes to car use that occurred during the fuel crisis. Statistical analyses (eg. correlations, regression models) may be helpful in exploring further the relationships within the data. The survey should prove to be a valuable resource of information for future travel behaviour analysis.

Discussion of the Results

Conditions during the Fuel Crisis

Before drawing conclusions regarding the implications of the survey results it is important to consider the nature of the situation faced by car users during the fuel crisis and how this relates to the conditions that they may experience in future. In effect the fuel crisis wielded a stick of unprecedented size against car use. At the same time the carrot of alternatives to the car did not change. It may have even got smaller in size because of service disruptions. In fact, Thorpe et al. (2002: Table 8.3) note in Chapter 8 that a small number of travellers switched from public transport to car during the fuel crisis. It is unlikely that any government would itself wield such a big stick in policy terms. Nevertheless, in the UK the stick against car use will increase in size with the introduction of policies such as road user charging. The government's ten year plan should increase the size of the carrot in terms of the availability and quality of alternatives to the car. So, the carrot and stick during the fuel crisis are of different size to those that exist normally or that might exist in future.

Another point to bear in mind is that conditions during the fuel crisis varied across the country and between individuals. In practice different parts of the UK suffered different levels of fuel shortage at different times and some individuals secured ample fuel for their immediate requirements, whilst others had little or none. Individuals would also have had different perceptions of the likely duration and severity of the situation.

There is perhaps an added complication. During the fuel crisis everyone faced an environment of fuel shortage and restrictions on car use. It is likely as a consequence that social norms were temporarily adjusted. Social norms are people's beliefs about the attitudes and behaviours that are normal, acceptable, or even expected in a particular social context. In many situations, people's perception of these norms will greatly influence their behaviour. During the fuel crisis it is likely that it became perceived amongst car users as much more acceptable to use other forms of transport. Travelling by bus may have not been considered a loss of social status during the fuel crisis because other people were having to use it. The survey showed for commuting that many respondents were happy to car share during the fuel crisis, although they would not wish to do so afterwards. When the crisis ended it is assumed that temporary distortions of social norms more or less disappeared.

Responses during the Fuel Crisis

The next question to ask is whether the circumstances of the fuel crisis will have produced responses of a similar magnitude and character as that which can be expected for future transport policy scenarios. An important characteristic of the fuel crisis is that there was no time for people to plan their responses to the circumstances created by the crisis or to adapt over time. Car users had to make tactical decisions on how to conserve fuel, whereas in response to policy measures they would be able to make strategic decisions. (This is discussed further in Chapter 14 (Bonsall, 2002)). In Section 6 of the survey it was asked whether respondents would reduce the number of car trips they make, move house to be nearer regular activities or buy a more efficient vehicle if fuel was rationed on a continuous basis. Of respondents 84 per cent said they would reduce car trips, 12 per cent said they would buy a more efficient vehicle and 4 per cent said they would move house. At face value, this suggests that most car users would not make long term changes and would instead make fewer trips by car.

There is a case to argue that the fuel crisis survey results represent a realistic indication of how car users would respond to policy measures promoting alternatives to the car and restricting car use. First, it should be reiterated that the survey captured actual responses to a real situation, unlike many surveys which obtain stated responses to a hypothetical situation. Second, those able to make changes to their car use, especially where they continued with their planned activity, were able to do so without much advance preparation. This indicates that their responses are feasible generally. Third, it has also been remarked previously that changes made during the fuel crisis had often been mentioned by respondents to have been made previously and as such are rehearsed alternatives.

Given the lack of time for car users to prepare responses to the fuel crisis it might even be argued that more substantial changes to car use would occur in future policy scenarios. There would be more time to examine alternative modes and destinations nearer to home. On the other hand, the trips cancelled in a crisis situation are unlikely to be dealt with in the same way in future policy scenarios. Other responses would be made, many of which would involve car use.

Having made these points, the large size of the stick on car use that existed during the fuel crisis has to be appreciated. The fuel crisis survey results are only likely to be indicative of policy scenarios which feature substantial restrictions on car use (e.g. road user charging).

Impacts of Car Reduction during Fuel Crisis

The following factors need to be considered when attempting to estimate the impacts on aggregate car travel of the changes made during the fuel crisis.

- *Frequency change made*: it has been noted in the survey whether changes were made on one occasion or more. Working from home was generally only carried out on one day, whilst children walking to school was generally carried out on at least two days.
- *Car travel remains involved in some changes*: for car sharing typically one car journey is removed for every two car sharing trips. Table 7.7 shows that train was used in combination with car (park and ride) in one-fifth of the instances train was used instead of car for the journey to work.
- *Travel distance*: for trip chaining the car is still used but there is usually a reduction in distance travelled and 'cold' starts.

Taking the example of school escort travel, about one-fifth of car users made their children walk to school. These respondents had an average journey length of 1.8 miles (see Table 7.8). Assuming the journey is not trip chained and they each did this for two days of the week they reduced their car travel by 7.2 miles each. For all school escort car users in the survey sample the average weekly mileage is 41 miles. The reduction of school escort car travel for the fuel crisis sample is calculated to be 4 per cent in terms of distance and 8 per cent in terms of trips. Table 7.1 shows that school escort car accounts for 4 per cent of all GB car trips and 2 per cent of all GB car distance travelled. The impact of this fuel crisis response is quite small at an aggregate level, although that it not to say that it is negligible or not important in other respects.

Insights on Car Dependence

The survey results have yielded insight regarding the character of car trips that are most amenable to change – those trips which have been labelled as marginal in a report on car dependence (Goodwin, 1995). The proportion of trips mentioned in that report (20 per cent) appears a reasonable estimate of the amount of car travel that can be categorised in this way.

The results indicate that alternative modes are presently viable for many people who usually use the car for commuting and school travel and indicate some aspects of these alternatives that need to be addressed to increase their use. For public transport, cost and travel time are the most important issues cited by respondents. Measures that reduce the inflexibility problems of car sharing will be needed to increase its use. Walking and cycling are recognised to be positive options but practical concerns (weather, carrying bulky items) are a barrier for their regular use. The use of local shops is hampered by the cost of products that they offer. Many of these findings are not new but they serve to reinforce existing strategies with hard evidence. The fact that some of the changes to car use are difficult to sustain on a regular basis (e.g. cycling, car sharing) should not discourage efforts to bring them about. People have more varied lifestyles than they used to and achieving change will require addressing many different options.

Comparison with other studies shows some consistency of results. Gärling et al. (2000) looked at the car use reduction measures perceived to be feasible by households in Gothenburg, Sweden. For commuting, public transport was the most likely choice. Closer destinations and trip chaining were identified for shopping and leisure trips. Kingham et al. (2001) examined what factors would influence employees in Hertfordshire to reduce car commuting and found potential for increased cycling and use of public transport if improvements in infrastructure and services were made. Mackett (2000) looked at measures required to reduce use of the car for journeys less than five miles. The survey sample were households in five areas of England. School escort journeys had the most potential for change. Of all short car trips, a switch to walking could be made for 31 per cent, to using the bus made for 31 per cent and to cycling made for 7 per cent. The measure that would have most impact on car use was found to be improving bus services.

The fuel crisis survey results also provide clear evidence that a majority of car travel at present is not considered to be viable to undertake by alternative means. Very little business and leisure travel was changed in any way other than simply not being undertaken. Table 7.1 indicates that leisure, business and other personal and escort travel accounts for 54 per cent of car trips and 65 per cent of all car distance travelled. Solutions to reducing car travel of this kind require efforts not only in transport but in land use planning and other policy areas.

Some survey respondents discussed how the fuel crisis had triggered them into considering sustained changes in travel behaviour:

> I bought a monthly season ticket…and intend using public transport from now on... Commuting by car from West Manchester to Halifax although usually fast is often stressful. Also driving with one person in the car is wasteful. (#602)

> Will use bus one day (a week). Work schedule changing and that will mean bus is more convenient than car. Might use bus another day if I find buses run to time ok. (#1605)

> This has hastened my arrangements to be able to work electronically from home more, which were being pushed by the rise in petrol price anyway. It's also given me more incentive to find another job nearer home, so I'm looking more seriously. (#1591)

These quotes highlight how events in people's lives, whether they are on a scale like the fuel crisis or personal events such as a new job, provide opportunities to reassess behaviour. Transport policy strategies should be developed acknowledging the opportunity of these occasions for creating behavioural change. The existence of these opportunities also brings some optimism to the prospects of bringing about change, as explained below:

> I would say that the only successful pathway to substantial change in transport behaviour at the aggregate level is by intervening to secure an 'asymmetric pattern of churn'. It means that we should stop talking in terms of encouraging people to stop driving and start using public transport – but seeking to increase a little the numbers of people who are already, every year, doing exactly that in huge numbers, and reducing a little the numbers of people who are already, every year, doing exactly the opposite, in equally huge numbers. (Goodwin, 1999)

Conclusions

The fuel crisis survey has shown that three-quarters of car users made a change to their car use during the fuel crisis. A lot of leisure, business and shopping travel was not undertaken. Of commuting car users 31 per cent changed mode or car shared. Of grocery shopping car users 14 per cent shopped more locally than usual, either going by car or walking or cycling.

Consideration of the special circumstances of the fuel crisis suggests the following. Where car use was reduced by not travelling this is not likely to be repeated in any future policy scenario. Where car use was reduced by using alternative modes or travelling shorter distances this gives a good indication of the absolute level of response and relative levels of response that would occur in a future scenario where transport policy acts mainly to restrict car use. The fuel crisis results, however, tell us less about the likely responses to policy that improves alternatives to the car.

The survey results provide reinforced evidence on areas fruitful to target in reducing car travel. These include fast and affordable public transport for journeys to work, car sharing and work from home programmes, school walking and cycling programmes and planning policy to promote affordable local food shops. The results also show that the majority of car travel is not amenable to change in the short term and will require sustained policy initiatives in transport and other policy areas if it is to be reduced. Policy action is required on two fronts: short term action to target marginal car travel and long term action to target entrenched car travel.

Events like the fuel crisis can not only reveal insight into behaviour but can trigger change. Very little of the change to car use that took place during the fuel crisis was probably sustained afterwards by those making it. This is not to say that events of various kinds should not be taken seriously as a means of influencing change. Finally, it should be remembered that bringing about change to a relatively small proportion of car use can make a substantial difference to congestion, pollution and other aspects of the quality of our lives.

References

Bhat, C.R. and Koppelman, F. S. (2000). Activity-Based Travel Demand Analysis: History, Results and Future Directions, *Preprint of paper presented at 79th Annual Meeting of Transportation Research Board*, Washington, D.C., 9-13 January.

Bonsall, P. (2002). What Can be Learned from Studying Transport Crises? In Lyons, G. and Chatterjee, K. (eds.) *Transport Lessons from the Fuel Tax Protests of 2000*, Ashgate, Aldershot.

Chatterjee, K. (2001). Asymmetric churn - academic jargon or a serious issue for transport planning? Paper prepared for Transport Planning Society.

DETR (2000). *Transport 2010 - The 10 Year Plan.* Department of the Environment, Transport and the Regions, TSO, London.

DETR (2001). *Transport Trends: 2001 Edition.* Department of the Environment, Transport and the Regions, TSO, London.

Gärling, T., Gärling, A. and Johansson, A. (2000). Household choices of car-use reduction measures. *Transportation Research Part A* 34(5), 309-320.

Goodwin, P. (ed) (1995). *Car Dependence.* RAC Foundation for Motoring and the Environment, London.

Goodwin, P. (1999). Action or Inertia? One Year on from 'A New Deal for Transport'. Transcript of Lecture given at Transport Planning Society Meeting at the Institution of Civil Engineers, *22 July 1999.*

ICM (2000a). The Guardian/ICM Monthly Poll - September 2000. Available at: http://www.icmresearch.co.uk/reviews/2000/guardian-poll-sept-2000.htm. [accessed 27/09/01].

ICM (2000b). BBC petrol crisis poll – September 2000. Available at: http://www.icmresearch.co.uk/reviews/2000/pertol-poll-september-2000.htm. [accessed 27/09/01].

Jones, P.M., Dix, M.C., Clarke, M.I. and Heggie, I.G. (1983). *Understanding Travel Behaviour*, Gower, Aldershot.

Kingham, S., Dickinson, J. and Copsey, S. (2001). Travelling to work: will people move out of their cars. *Transport Policy* 8(2), 151-160.

LFS (2001). SOC 2000 and NS-SEC on the Labour Force Survey. Available at: http://www.statistics.gov.uk/themes/labour_market/articles/socchanges.asp. [accessed 27/09/01].

Lex (1999). *The 1999 Lex Report on Motoring (11th Edition).* Lex Service, Bourne End, Bucks.

Lyons, G. and Beecroft, M. (2002). Experiences and Lessons From Conducting an Emergency Survey of Motorists During the UK Fuel Crisis. In Lyons, G. and Chatterjee, K. (eds.) *Transport Lessons from the Fuel Tax Protests of 2000*, Ashgate, Aldershot.

Mackett, R.L. (2000). How to Reduce the Number of Short Trips by Car. In *Proceedings of Seminar C of the European Transport Conference, Homerton College, Cambridge, 11-13 September*, pp. 177-190, P438, PTRC Education and Reseach Services Ltd, London.

ONS (2000). *Standard Occupational Classification, 2000.* Great Britain Office for National Statistics, TSO, London.

OPCS (1992). *1991 Census: Preliminary Report for England and Wales*, Office of Population Censuses and Surveys, London.

Thorpe N., Bell, M., Polak J. and Noland R. (2002). A Telephone Survey of Stated Travel Responses to Fuel Shortages. In Lyons, G. and Chatterjee, K. (eds.) *Transport Lessons from the Fuel Tax Protests of 2000*, Ashgate, Aldershot.

TYPE OF CHANGE MADE TO CAR USE: MASTER-LIST

No.	Type of change	Group[a]
1	Bus	5
2	Train	4
3	Bus and train	4
4	Car and bus	5
5	Car and train	4
6	Walked car part of journey	7
7	Walk and Bus	5
8	Walk and Train	4
9	Cycled car part of journey	6
10	Cycle and Train	4
11	Used hire or pool car	8
12	Used different household car	8
13	Car/bus[2]	5
14	Car/train[2]	4
16	Used bus in car part of journey	5
17	Motorcycle	8
18	Walked/Car (got lift)[b]	7
19	Bus/Train[b]	4
20	Cycle and car	6
21	Walked	7
22	Cycled	6
23	Took taxi	8
24	Taxi and train	4
25	Bus/taxi[b]	5
26	Worked in nearer office going by foot	2
27	Worked in nearer office going by bicycle	2
28	Car alone	8
31	Car shared (got lift)	9
32	Car shared (gave lift)	9
33	Car shared (not sure if got or gave lift)	9
34	Car shared on rota basis	9
35	Altered car sharing arrangement	9
36	Motorcycle (got lift)	9
37	Lift (by car) for car part of journey	9
41	Worked from home	1
42	Cancelled trip(s)[c]	1
43	Postponed trip(s) [c]	1
44	Cancelled trip as work/school/shop closed	1
45	Someone else made trip(s) instead	1
46	Not undertaken because linked to other trip that was cancelled	1

TYPE OF CHANGE MADE TO CAR USE: MASTER-LIST

No.	Type of change	Group[a]
47	Had enough food and made do	1
51	Trips more locally by car[d]	2
52	Consolidated trips (linked them together into tour)	3
53	Linked to other car trip (not specified)	3
54	One large shopping trip by car	1
55	Used shops local to home by foot	2
56	Used shops local to home by car	2
57	Linked trip to morning car commute trip	3
58	Linked trip to evening car commute trip	3
59	Linked trip to car shopping trip	3
60	Linked trip to car children-to-school trip	3
61	Made priority trips only	1
62	Linked activity to other public transport trip	2
63	Walked to shop near workplace	2
64	Went directly to work/shop instead of trip chaining	3
65	Used shops local to home by bicycle	2
66	Worked in nearer office with mode not indicated	2
67	Made extra trips to complete shopping	11
68	Used shops at work, unknown mode	2
69	Worked in nearer office taking bus	2
70	Consolidated trips (did not go home and return)	1
71	Drove more fuel efficiently	10
72	Changed route driven to avoid petrol queues	13
73	Changed route/time driven to avoid demonstrators	13
74	Someone else got groceries for me/us	1
75	Stayed overnight at work-place/friend	2
76	Changed to more fuel efficient route	10
77	Got more groceries than usual	13
78	Made trip earlier in week	12
79	Changed route to find fuel	13
80	Changed route due to traffic better	13
81	Changed travel time to save fuel	10
82	Could not get some shopping	13
83	Change not related to fuel crisis	13
84	Travel as normal but destination activity not possible due to fuel crisis	13

a Type of change group: 1=trip reduction, 2=trip origin or destination change, 3=trip chaining, 4=train, 5=bus, 6=cycle, 7=walk, 8=other vehicle, 9=car share, 10=driving style, 11=trip increase, 12=trip rescheduling for earlier, 13=other.

b First named mode used out and second named mode used return or vica versa.

c May be due to difficulties of travel for other person.

d For commuter who has fixed normal work-place this means going to more local office.

8 A Telephone Survey of Stated Travel Responses to Fuel Shortages

NEIL THORPE, MICHAEL BELL, JOHN POLAK AND
ROBERT B. NOLAND

Introduction

The objective of this study was to use the fuel crisis to investigate how travel adapts to the progressive removal of the car and, as a consequence, to throw light on the barriers to a shift to more sustainable modes of transport. Short-term adaptations could range from changing mode, reducing the frequency of short distance trips, cancelling activities, substituting travel for information technology, to trip chaining. Long-term adaptations could include moving home or job. Barriers to adaptation could result from a lack of information, a perceived (rather than real) lack of alternatives, or difficulties in trip planning. A shift to public transport could be limited by the geographic coverage of the services or difficulties in using buses, in particular for those encumbered with luggage or some other form of mobility handicap. One consequence of a reduction in mobility could be a reduced range of activities, both in time and space.

To investigate these issues, three focus groups were organised followed by 1001 telephone interviews, representing urban, conurban and rural areas of the UK (Polak et al., 2001). This chapter presents the main findings from the analysis of the quantitative data from the telephone survey about the potential for modal substitution, for different trip purposes and for three regions of the country. A complex picture of modal dependency emerges, with travel choices constrained by physical and informational barriers, and influenced by life cycle, gender and activity-specific factors. Although these findings relate to the extreme scenario of a fuel shortage, they indicate the kinds of adaptations that would have to accompany any significant traffic reduction. The implications of the findings for transport policy are developed in the conclusions.

The course of the fuel protest has been described as follows:

> The first rumblings ... were heard on 5 September when it was announced that fuel prices were to rise again following a rise in the price of crude oil. The Channel Tunnel was blockaded in protest on 6 September and on 7 September the first oil refinery, at Stanlow, Cheshire, was blockaded. The protest spread very rapidly. More refineries were blockaded on the 8 September and on 9 September nation-wide panic-buying began. On Sunday 10 September protests had closed Britain's largest inland oil terminal at Kingsbury, West Midlands, and huge queues at garages were reported. By Tuesday 12 September protesters had blocked six of the UK's eight refineries and over half of Britain's filling stations were shut.
>
> The protest ended almost as quickly as it had begun. On Thursday 14 September the Stanlow blockade ended and on Friday 15 September the first deliveries were reaching some garages, although it was estimated that 90 per cent of filling stations were empty of fuel. (Hathaway, 2000)

The week-long blockade of the refineries in September 2000 led to a massive transfer of fuel from filling stations to vehicles, causing filling stations to run dry. Individual and corporate vehicle users were forced to reassess their travel plans, with the result that car traffic fell significantly. On 14 September, car traffic (as measured by the (then) Department of the Environment, Transport and the Regions' Automatic Traffic Data Collection system at 135 sites throughout Great Britain) was lower than average by 39 per cent on motorways, 25 per cent on major roads and 16 per cent on minor roads (Hathaway, 2000). Road traffic in the third quarter of 2000 was reduced by nearly 2 per cent as a result of the fuel protest, with the result that the estimate of total traffic in 2000 is about 0.4 per cent lower than it would otherwise have been. Goods traffic seems to have been less affected. On 14 September, goods traffic was lower than average by 13 per cent on motorways, 13 per cent on major roads and 9 per cent on minor roads (Hathaway, 2000). Nonetheless, the impact of the protest rippled through the national economy, affecting retailing, just-in-time manufacturing, warehouses, hospitals, tourism, schools, the Post Office, taxis and bus operators.

Aims and Objectives

The fuel crisis restricted car use and forced necessary adaptations in travel behaviour. These adaptations are potentially of a nature and scale not normally achieved by conventional transport policy measures (including high fuel excise duties). The incidence of these coerced responses provided a unique opportunity to observe the mechanisms of adaptation involved in major reductions in car use and to gain insight into the consequences of these adaptations for the individuals concerned and their families, both in the short- and long-term.

The aim of the study was therefore to take advantage of this opportunity by undertaking qualitative and quantitative empirical research to address the following questions:

- What aspects of existing travel and activity patterns are affected by a coerced reduction in car use (e.g. which purposes and times of day are most seriously affected, are effects confined to travel or do they spill over into time use and activities)?
- Which adaptations are made to travel behaviour and activity patterns to accommodate these reductions (e.g. are adaptations confined just to changes of mode or do they also include switching destination, dropping, combining or re-timing certain activities, vehicle sharing, switching expenditure into travel from other areas to support the use of taxis)?
- What are the mechanisms leading to the chosen adaptations (e.g. how many alternatives are considered, is new information acquired, are other household members/work colleagues consulted/affected)?
- What difficulties are encountered in planning and executing travel adaptation strategies (e.g. incompatibilities/rigidities of public transport schedules, lack of information, interchange difficulties, heavy loads)?
- What are the limits to adaptations (e.g. to what extent is the car indispensable and how sustainable are the chosen adaptations in the short- and long-term)?

One interesting hypothesis is that crises force technological change, as in the case of the widespread use of the fax, which resulted from the postal disputes of the 1980s. It could be that the fuel crisis produced a decisive shift toward the use of telecommunications and the Internet for working and shopping or toward the use of more fuel-efficient vehicles.

Methodological Approach to the Telephone Survey

A telephone survey was commissioned by the authors to investigate, amongst other things, the potential adaptations to car removal and the perceived barriers to adaptation. MORI conducted 1001 telephone interviews, filtered by car access and selected randomly from across London, the South East and the North East of England. To allow a geographic analysis of the data, the first part of the postcode was recorded. The interview, which lasted about 8 minutes, had two parts; the first related to actual responses to the fuel crisis, while the second related to stated responses to a hypothetical fuel shortage sometime in the future. Due to particular resource constraints, telephone surveys were considered the most appropriate approach in terms of delivering sufficient numbers of respondents in the sample areas, within the short space of time available for data collection, and while experiences of the fuel crisis were still fresh within the minds of respondents. It is worth noting that this approach tends to elicit individuals' (rather than households') immediate responses to questions regarding their behaviour.

Sample Composition and Evidence of Mode Shift During the Crisis

Of those sampled, 29 per cent ran out of fuel during the fuel crisis or had insufficient fuel to make their normal journeys. Only 15 per cent had access to emergency fuel supplies. The majority of respondents (71 per cent) were working, 4 per cent were either attending school or college, and the remainder (25 per cent) did neither. Of respondents 55 per cent had access to one car, 37 per cent to two cars, 7 per cent to three cars and 1 per cent to four or more cars. Of respondents 6 per cent lived in households with access to a motorcycle. The age distribution of respondents, shown in Table 8.1, shows that just under half (46 per cent) were in the 35 to 54 age group.

Table 8.2 shows changes in mode choice before and during the fuel crisis for travellers to work, school or college. Aside from those who chose to work from home, the biggest proportionate increase is for car as passenger (63 per cent) whereas the biggest absolute increase is for public transport (35 respondents). The nature of these modal shifts is analysed in Table 8.3. Of 582 respondents who chose car as driver before the fuel crisis, 422 (73 per cent) remained with the car as driver, 50 (9 per cent) chose public transport, 30 (5 per cent) walked and 20 (3 per cent) went by car as a passenger. Interestingly, of 88 respondents who went by public transport before the crisis, 9 (10 per cent) switched to car as driver, so the proportionate switch between these two modes is approximately equal. The switch from public transport to car as driver could be the combined result of over-crowding on public transport and reduced congestion levels on the roads.

Table 8.1 Age distribution of respondents

Age Group (years)	Number of respondents	% of Respondents
16-24	82	8
25-34	223	22
35-54	459	46
55 and above	233	23
No response	4	1
TOTAL	1001	100

For London respondents only, the shift from car as driver to public transport is proportionately larger (at 19 per cent), but to car as passenger or walking is proportionately smaller (at 2 per cent and 4 per cent respectively). This is probably due to the better provision of public transport and the longer commuting distances in London. Of respondents in London 7 per cent switched from public transport to car as driver, slightly lower than the corresponding figure for the sample as a whole. The picture for the North East however is rather different. The shift from car as driver to public transport is much smaller (at 4 per cent), while the shift to car as passenger or walking is also small (at 2 per cent each), although the small size of the sub-sample makes strong conclusions impossible.

Of those respondents who continued to drive to work, school or college during the fuel crisis, 402 used the same car and 24 used a more fuel-efficient car, possibly reflecting the limited scope in practice for people to swap between alternative vehicles at short notice.

Table 8.2 Choice of main mode for journeys to work/school/college before and during the fuel crisis

Mode of travel	Number of respondents using this mode	Number of respondents using this mode during the fuel crisis	% change
Car as driver	563	429	-24
Car as passenger	22	36	+63
Motorcycle	4	4	0
Public transport	84	119	+42
Bicycle	19	29	+53
Walk	73	100	+37
Van/truck	5	4	-20
Other	16	5	-69
Work(ed) from home	3	19	+633
Don't know	0	1	-
Total	789*	746	-

* 33 respondents provided more than one answer

Of the respondents who normally drive to work, school or college but did not do so during the fuel crisis, 38 (or 33 per cent) reported no difficulty in using another mode, 34 (or 28 per cent) had major difficulties in using another mode, 26 (or 22 per cent) reported minor difficulties in using another mode, and 18 (or 15 per cent) said it was impossible to use another mode and so did not travel to work, school or college during this period.

Table 8.3 Mode shifts for journeys to work/school/college during the fuel crisis (all respondents)

Before/ During	Car as driver	Car as passenger	Motorcycle	Public transport	Bicycle	Walk	Van or truck	Other	Work from home	Don't know	Did/could not get to work	Unaffected as on holiday	Unemployed during the crisis	TOTAL (before)
Car as Driver	422	20	2	50	13	30	0	5	7	1	12	16	4	582
Car as passenger	1	15	0	4	0	5	0	0	0	0	0	0	0	25
Motor-cycle	1	1	4	0	0	0	0	0	0	0	0	0	0	6
Public transport	9	0	0	73	1	4	0	0	0	0	0	1	0	88
Bicycle	2	0	0	1	16	2	0	0	0	0	0	0	0	21
Walk	3	2	0	3	1	68	0	0	0	0	0	1	1	79
Van or truck	0	0	0	0	0	0	4	0	0	0	1	0	0	5
Other	0	2	0	0	0	1	0	0	0	0	0	0	0	3
Work from home	1	0	0	1	0	1	0	1	12	0	0	0	0	16
Don't know	0	0	0	0	0	0	0	0	0	0	0	0	0	0
TOTAL (during)	439	40	6	132	31	111	4	6	19	1	13	18	5	825

Disruption to Travel Patterns

Table 8.4 shows respondents' expectations of disruption caused by a hypothetical fuel crisis tomorrow. ('We would now like you to consider the impacts on your travel if fuel supplies were to run out tomorrow with some uncertainty as to when your petrol station would have new supplies available...please rate how the following types of journey might be affected.') In the case of commuting, travel as part of work, or visiting friends and family, approximately equal proportions of respondents expect and do not expect to be disrupted. In the case of child centred activities (such as taking children to school), going to the shops, indulging in hobbies and pastimes, or visiting a doctor, the majority of respondents do not expect disruption, presumably because they feel they can use public transport or go on foot.

Correlations in disruption may be indicative of trip chaining, the interaction of demographic variables or other (unmeasured) factors. Table 8.5 shows that disruption to commuting associates with disruptions to travel in work, possibly because the same car is often used for both purposes. Disruption to the school-run associates with disruption to trips to out-of-school activities, again because the same car may be frequently used for both purposes.

Stated Behavioural Adaptations

Tables 8.6 to 8.10 provide further information regarding the stated behavioural responses to questions relating to a hypothetical fuel crisis tomorrow. There is strong disagreement from a majority of respondents with the proposition that working at home would offer a suitable alternative to travelling to work, reflecting the nature of most jobs, which are tied to a work place. The majority consider it possible to travel to work by public transport, walking or cycling, although almost a third of the sample were of the opposite opinion. While a majority felt that they could shop for food on foot, there was a clear aversion by many to shopping by public transport. A majority felt they would not reduce the frequency with which they shopped for food, perhaps reflecting a requirement for fresh food. The respondents were divided on whether the Internet offered a suitable alternative for shopping, whether they could walk their children to school, or whether they could engage in recreational activities closer to home. Most respondents were opposed to ride sharing with another family member or friend. Although they were divided on whether better planning could save car trips, a majority felt that with better planning they would be able to use public transport more frequently.

The responses to these questions were then filtered for those expecting disruption to trips of different types on the basis that those expecting disruption would be more likely to make behavioural adaptations. The results were considered for the sample as a whole, as well as separately for London and the North East. In many cases, however, the filtering led to sub-sample sizes for the North East that were too small to draw firm conclusions.

Of those expecting disruption to any kind of trip in a future fuel crisis, Table 8.7 confirms the (perhaps not surprising) result that for most people working at home is not an option. This result applies to London and the North East as well. The divided response in Table 8.8 suggests that for those expecting to be disrupted, public transport, walking or cycling offer a convenient way to get to work for only some of the sample.

Table 8.9 reveals that working at home is not an option for the majority of those expecting disruption to commuting in a future fuel crisis. However, the sample is more divided about whether public transport, walking or cycling offers a viable alternative, and ride sharing is not an option for the majority. Surprisingly many (nearly 25 per cent) are unsure whether better personal logistics would save car trips. Finally, a majority feel that fuel-efficient vehicles would not be an option for them. However, as the question related to a fuel crisis tomorrow, this cannot be taken as an indication that the majority feel that more fuel efficient vehicles will not be an option for them at some future time. The results for London are broadly similar, but in the North East respondents are divided (rather than unsure) about the potential for saving car trips through better personal logistics.

For those expecting significant disruption to the school run, Table 8.10 shows that the majority could not walk children to school, although there was a sizeable group (41 per cent) who felt they could. Ride sharing is not an option for the majority. For the majority, better planning would not save car trips but might facilitate more public transport use. The majority could not, in the short-term, use more fuel-efficient vehicles.

For those expecting disruption to shopping trips, Table 8.11 shows a large proportion of the respondents (55 per cent) are able to shop on foot compared to 39 per cent who are not. Public transport however is not considered by many to be a viable alternative. Respondents would do their shopping less frequently in 33 per cent of cases and even fewer (19 per cent) would use the Internet. Ride sharing with another family member or friend is not a popular alternative. There was uncertainty about whether better planning could lead to fewer car trips. Using a more fuel-efficient vehicle would not be an option for most. The results for Londoners were broadly similar, but they seemed to be less sanguine about the possibility of saving car trips by better planning, presumably as they already face more obstacles to car use (such as parking availability and cost).

For those expecting disruption to visits of friends and family, Table 8.12 suggests that ride sharing would not be an option for most. There was uncertainty about whether car trips could be saved through better planning. Few felt they would be able to use a more fuel-efficient vehicle. In London, ride sharing was favoured by relatively more respondents. Ride sharing was not favoured in the North East.

For those expecting disruption to trips for hobbies and pastimes, Table 8.13 indicates a division of opinion over whether opportunities closer to home could be used. Ride sharing is unpopular for these trips. There is uncertainty from many respondents about the scope for saving car trips through better planning. Using a more fuel-efficient vehicle is again not a widespread option. Londoners were less negative about ride sharing.

Table 8.4 Perceived impact of future fuel shortages on different types of trip

Trip type	Not affected 1	2	3	Severe disruption 4	5	Total
Commuting to/from work, school or college	249 *(29.9%)*	88 *(10.6%)*	102 *(12.2%)*	102 *(12.2%)*	292 *(35.1%)*	833 *(100%)*
Travelling as part of work	261 *(36.5%)*	50 *(7%)*	61 *(8.5%)*	80 *(11.2%)*	264 *(36.8%)*	716 *(100%)*
Taking children to/from school	325 *(57.8%)*	45 *(8%)*	55 *(9.8%)*	38 *(6.7%)*	100 *(17.7%)*	563 *(100%)*
Going to the shops	426 *(43.4%)*	160 *(16.3%)*	170 *(17.3%)*	96 *(9.8%)*	130 *(13.2%)*	982 *(100%)*
Visiting friends and family	242 *(24.4%)*	50 *(15.1%)*	208 *(21%)*	150 *(15.1%)*	241 *(24.3%)*	991 *(100%)*
Hobbies and past-times	391 *(40.2%)*	155 *(15.9%)*	160 *(16.4%)*	115 *(11.8%)*	152 *(15.6%)*	973 *(100%)*
Taking children to out-of-school activities	263 *(43.1%)*	79 *(13%)*	96 *(15.7%)*	62 *(10.2%)*	110 *(18%)*	610 *(100%)*
Trips to the doctor or health-care provider	517 *(52.5%)*	122 *(12.4%)*	111 *(11.3%)*	80 *(8.1%)*	155 *(15.7%)*	985 *(100%)*

Table 8.5 Correlation of anticipated disruption to different trip types

Trip Purpose	Commuting	Travel as part of work	School-run	Shopping	Visiting friends and family	Hobbies and past-times	Taking children to out-of-school activities	Visits to the doctor/ health-care provider
Commuting	-	0.55	0.29	0.17	0.18	0.2	0.24	0.23
Travel as part of work	-	-	0.27	0.19	0.18	0.19	0.23	0.22
School-run	-	-	-	0.29	0.21	0.23	0.57	0.39
Shopping	-	-	-	-	0.36	0.27	0.25	0.39
Visiting friends and family	-	-	-	-	-	0.38	0.31	0.27
Hobbies and past-times	-	-	-	-	-	-	0.33	0.22
Taking children to out-of-school activities	-	-	-	-	-	-	-	0.28
Visits to the doctor/health-care provider	-	-	-	-	-	-	-	-

Table 8.6 Possible behavioural adaptations to future fuel shortages

Behavioural Response	Strongly agree 1	2	3	4	Strongly disagree 5	Total
'I could work from home on a frequent and regular basis'	96 *(10.7%)*	87 *(9.7%)*	80 *(8.9%)*	112 *(12.5%)*	522 *(58.2%)*	897 *(100%)*
'I could easily travel to work by public transport, walking or bicycling'	267 *(28.5%)*	327 *(34.9%)*	46 *(4.9%)*	104 *(11.1%)*	192 *(20.5%)*	936 *(100%)*
'I could do my food shopping on foot'	257 *(26.5%)*	345 *(35.6%)*	40 *(4.1%)*	121 *(12.5%)*	205 *(21.2%)*	968 *(100%)*
'I could use public transport to do food shopping'	109 *(11.2%)*	122 *(12.6%)*	42 *(4.3%)*	199 *(20.5%)*	98 *(51.3%)*	970 *(100%)*
'I could do my shopping less frequently'	212 *(24.2%)*	161 *(18.4%)*	88 *(10.1%)*	78 *(8.9%)*	336 *(38.4%)*	875 *(100%)*
'I could do some of my food shopping via the Internet'	289 *(29.4%)*	244 *(24.8%)*	27 *(2.7%)*	142 *(14.5%)*	280 *(28.5%)*	982 *(100%)*
'I could walk my child to school'	232 *(23.6%)*	276 *(28%)*	38 *(3.9%)*	160 *(16.2%)*	279 *(28.3%)*	985 *(100%)*
'I could engage in recreational activities near my home'	219 *(22.4%)*	332 *(34%)*	51 *(5.2%)*	170 *(17.4%)*	204 *(20.9%)*	976 *(100%)*
'I could easily share a ride with another family member or friend for many of my trips'	96 *(9.8%)*	204 *(20.7%)*	34 *(3.5%)*	134 *(13.6%)*	516 *(52.4%)*	984 *(100%)*
'I could plan my trips better so that I combine trips and don't make unnecessary car trips'	170 *(24.3%)*	125 *(17.9%)*	194 *(27.7%)*	34 *(4.9%)*	177 *(25.3%)*	700 *(100%)*
'I could use another motor vehicle that has better fuel efficiency'	209 (21.3%)	228 (23.2%)	44 (4.5%)	181 (18.4%)	320 (32.6%)	982 (100%)

Table 8.7 Responses to statement 'I could work from home on a frequent and regular basis' (number of respondents)

	Strongly Agree	Tend to Agree	Neither Agree nor Disagree	Tend to Disagree	Strongly Disagree	Already Do This	Don't Know	Total
Remainder of sample	27	29	12	54	258	10	4	394
London	8	7	0	11	42	4	0	72
Rest of South East	5	8	6	15	71	1	1	107
North East	2	1	0	3	20	0	0	26

Table 8.8 Responses to statement 'I could easily travel to work by public transport, walking or cycling' (number of respondents)

	Strongly Agree	Tend to Agree	Neither Agree nor Disagree	Tend to Disagree	Strongly Disagree	Already Do This	Don't Know	Total
Remainder of sample	78	117	16	55	106	19	3	394
London	11	21	1	11	23	4	1	72
Rest of South East	23	30	6	13	29	6	0	107
North East	4	8	1	3	9	1	0	26

Table 8.9 Responses to statements about behavioural adaptations to expected disruption to commuting trips (number of respondents)

	Strongly Agree	Tend to Agree	Neither Agree nor Disagree	Tend to Disagree	Strongly Disagree	Already Do This	Don't Know	Total
'I could work at home on a frequent and regular basis'	27	29	12	54	258	10	4	394
'I could easily travel to work by public transport, walking, or bicycling'	78	117	16	55	106	19	3	394
'I could easily share a ride with another family member or friend for many of my trips'	36	92	15	50	195	2	4	394
'I could plan my trips better so that I combine trips and don't take unnecessary car trips'	61	57	59	18	78	28	93	394
'I could use another motor vehicle that has better fuel efficiency'	53	69	11	85	167	6	3	394

Table 8.10 Responses to statements about behavioural adaptations to expected disruption to the school run (number of respondents)

	Strongly Agree	Tend to Agree	Neither Agree nor Disagree	Tend to Disagree	Strongly Disagree	Already Do This	Don't Know	Total
I could walk my child to school	21	35	5	23	49	0	0	138
I could easily share a ride with another family member or friend for many of my trips	16	26	6	22	66	0	2	138
I could plan my trips better so that I combine trips and don't take unnecessary car trips	24	23	6	15	66	1	3	138
I could use another motor vehicle that has better fuel efficiency	18	23	8	29	56	2	2	138

Table 8.11 Responses to statements about behavioural adaptations to expected disruption to the shopping trips (number of respondents)

	Strongly Agree	Tend to Agree	Neither Agree nor Disagree	Tend to Disagree	Strongly Disagree	Already Do This	Don't Know	Total
I could do my food shopping on foot	44	80	8	32	56	5	1	226
I could use public transport to do food shopping	22	26	11	41	114	7	5	226
I could do my food shopping less frequently	39	36	18	14	89	13	17	226
I could do some of my shopping via the internet	15	29	8	47	126	1	0	226
I could easily share a ride with another family member or friend for many of my trips	20	42	8	28	126	1	1	226
I could plan my trips better so that I combine trips and don't take unnecessary car trips	31	27	46	10	50	9	53	226
I could use another motor vehicle that has better fuel efficiency	24	33	13	48	103	3	2	226

Table 8.12 Responses to statements about behavioural adaptations to expected disruption to visits to friends and family (number of respondents)

	Strongly Agree	Tend to Agree	Neither Agree nor Disagree	Tend to Disagree	Strongly Disagree	Already Do This	Don't Know	Total
I could easily share a ride with another family member or friend for many of my trips	29	90	19	43	206	2	2	391
I could plan my trips better so that I combine trips and don't take unnecessary car trips	65	44	81	16	71	12	102	391
I could use another motor vehicle that has better fuel efficiency	50	87	17	70	158	5	4	391

Table 8.13 Responses to statements about behavioural adaptations to expected disruption to trips for hobbies and past-times (number of respondents)

	Strongly Agree	Tend to Agree	Neither Agree nor Disagree	Tend to Disagree	Strongly Disagree	Already Do This	Don't Know	Total
I could engage in recreational activities near my home	56	77	10	52	64	7	1	267
I could easily share a ride with another family member or friend for many of my trips	24	63	10	33	134	2	1	267
I could plan my trips better so that I combine trips and don't take unnecessary car trips	40	26	60	10	50	14	67	267
I could use another motor vehicle that has better fuel efficiency	39	55	12	52	102	6	1	267

Conclusions

The fuel crisis of September 2000 provided a unique opportunity to study adaptation to the progressive removal of the car thereby providing evidence of the barriers to the use of more sustainable modes of transport. The following are some of the main conclusions and policy recommendations emerging from the qualitative and quantitative investigations.

Working at home is only an option for certain people some of the time. Evidence from the focus groups suggested this, and data from the telephone survey confirmed it. Public transport, walking or cycling offers a viable alternative to commuting by car for a large number of people. With better planning, more use could be made of public transport. However, many regard public transport as unreliable, slow and unsuited to certain types of trip (like food shopping). Policy could focus on improving the quality and reliability of public transport as well as trip planning and travel information services. The telephone survey suggests that proportionately fewer car commuters in London regard public transport as an option, probably due to the increased hurdles already faced by drivers in the capital.

Ride sharing is not a popular option. However, the telephone sample suggested a lower aversion to ride sharing in London than elsewhere, perhaps because fewer regarded public transport as an option, higher residential density and longer commutes. Surprisingly many respondents are unsure whether they could save car trips with better planning. This points to a niche for improved personal logistics, involving better trip chaining and/or destination choices. Research in this area would be welcome, perhaps harnessing the power of networked electronic organisers.

A switch to more fuel-efficient vehicles is not a short-term option, but may be part of the long-term solution. Indeed, the focus groups suggested an awareness of environmental issues, particularly via the young at school. It is accepted that policy should be directed to encouraging the use of more environmentally friendly fuels and more fuel-efficient vehicles. There was also a belief that private transport was here to stay and that fossil fuels will gradually be replaced by alternatives.

Many respondents felt able to shop on foot. The focus groups confirmed that conventional bus travel is not a popular alternative for shopping, because of difficulty of use for the encumbered, the inflexibility of the route, lack of reliability and cost. Policy could be directed toward achieving greater reliability and improved access through the use of low floor vehicles. There appears to be a market for demand responsive services, delivering greater route flexibility with higher reliability using smaller, more accessible vehicles. The logistics behind demand responsive services can now be solved by a combination of modern vehicles, telecommunications and logistics software.

Few would do their shopping less frequently, presumably because of the need for fresh food. The telephone survey indicates that even fewer would use the Internet for shopping. The focus groups suggest that this is due to a desire for social interaction as well as a wish to see and touch the produce. There is little evidence in the qualitative or quantitative data that the fuel crisis significantly advanced the use of the Internet for either shopping or work.

Acknowledgements

The authors would like to acknowledge the kind support from Transport for London and the Commission for Integrated Transport who funded this research, MORI who conducted the telephone interviews and Social and Transport Research Services Ltd. who were responsible for the focus group activities.

References

Hathaway, P. (2000). *The effect of the fuel protest on road traffic*. Transport Statistics Road Traffic Division, DETR.

Polak J.W., Noland R.B., Bell M.G.H., Thorpe, N. and Wofinden, D. (2001). What happens when the pumps run dry? Experience from the 2000 fuel crisis and its policy implications. Paper presented at the European Transport Conference Seminar on Transport Management, Cambridge (September).

9 Stages of Change in Drivers' Willingness to Reduce Car Use, Before and During the UK Fuel Crisis

SUSAN BEATTY, MICHELLE MEADOWS AND DAVID WHITE

Introduction

Between 1988 and 1997, car ownership in Britain rose by over 20 per cent, to 25.8 million cars, an average of one car for every 2.2 people (Lex Report on Motoring, 1998). About 72 per cent of British homes own at least one car and 27 per cent of households have two or more cars (DETR, 2000). It is forecast that over the next few years the number of households with more than one car will continue to increase (Lex Report on Motoring, 1998).

The Government's White Paper on the future of transport (DETR, 1998) acknowledged the importance of the car to millions of people. At the same time, it recognised the fact that dependence on the car should be reduced and that the increasing congestion and pollution caused by higher volumes of traffic should be addressed urgently. The White Paper set out a strategy for creating more sustainable and integrated transport systems by improving public transport and by making it easier for people to walk or cycle. It pointed out that local authorities, public transport operators, employees and individuals all have their part to play in effecting change.

The fuel shortages of September 2000, and the political developments since then, served to highlight the extent of many people's reliance on their cars. The prospect of being unable to use their cars because of fuel shortages caused widespread alarm, manifested in long queues at garages and panic buying of fuel. The knock-on effects of the fuel shortages, for instance affecting food supplies in shops and the ability of hospital staff to get to work, further emphasised the extent of the country's dependency on the car.

The focus of this chapter is the application of the transtheoretical model (Prochaska and DiClemente, 1982) to drivers' willingness to reduce their car use, as measured during April 2000 and again in September 2000. An introduction to the transtheoretical model is provided, followed by a brief summary of the pre-fuel crisis study undertaken in April. The September fuel crisis study is then introduced, the results are described and conclusions drawn with reference to the earlier study.

The Transtheoretical Model

Research undertaken at Staffordshire University (Stoke-on-Trent and Stafford) during April 2000 (Beatty, 2001) involved a questionnaire being sent to 1612 University employees in order to measure their willingness to reduce car use. A response rate of 54 per cent resulted in 872 questionnaires being returned, of which 855 were used in analysis. The questionnaire used the transtheoretical model (Prochaska and DiClemente, 1982) to provide a framework within which to study drivers' car use. This model regards people as being in one of several stages (commonly 5) in relation to making a behavioural change, the earliest stage being precontemplation (not yet thinking about changing) and the latest stage being maintenance (when all possible change has been effected). According to the behaviour being studied, it may also be appropriate to consider an additional termination stage, in which all temptation to return to the old behaviour has ceased.

Other measures are commonly associated with the model; these measures are known as decisional balance, processes of change and self-efficacy. Decisional balance involves weighing up the pros and cons of making a behavioural change. Those in precontemplation tend to report that the cons of changing outweigh the pros. However, for those in the later stages, the balance alters so that the pros of changing outweigh the cons. In a study of 12 different behaviours, Prochaska et al. (1994b) found that the crossover point of pros and cons typically occurred during contemplation, although in the case of 5 behaviours it occurred at a later stage.

Processes of change are activities which involve modifications to a person's actions, emotions, relationships and thoughts (Holz, 1996). Examples of these processes are consciousness raising (increasing one's level of awareness of the behaviour by accessing information from different sources), dramatic relief (experiencing an emotional response, such as anger, towards the behaviour), reinforcement management (rewarding oneself for a behavioural change) and helping relationships (enlisting other people's support in trying to change the behaviour). It has been found that people at different stages of change use the processes of change to greater or lesser extents and also that at any one stage certain processes may be used more or less than others (e.g. Fava et al., 1995; Perz et al., 1996; Prochaska et al., 1991).

Self-efficacy (Bandura, 1977) is one's belief in being able to achieve a particular outcome, or perform a specific task, particularly in the face of certain difficulties. It tends to increase as people progress through the stages of change in the model (Holz, 1996).

The model was originally applied to the study of addictive behaviours, particularly smoking (e.g. Fava et al., 1995; Prochaska and DiClemente, 1983). It has more recently been applied to the study of a range of other behaviours including exercise take-up (e.g. Cardinal and Sachs, 1996; Clarke and Eves, 1997) and dietary change (e.g. Brug et al., 1997; Povey et al., 1999). It has been suggested as a potentially useful model for studying travel behaviour (Ferguson et al., 1999) and has been applied to a limited extent in this area (e.g. Mutrie et al., 1999).

The Pre-Fuel Crisis Study

The questionnaire used by Beatty (2001) measured drivers' willingness to reduce their car use by participants allocating themselves to one of five stages of change by choosing the description which best fitted their situation. The five stages were:

- *precontemplation* ('I have not tried to reduce the amount I use my car over the last 12 months, and I am not thinking of doing so over the next 12 months'); 41 per cent of drivers were in this stage.
- *contemplation* ('I have not tried to reduce the amount I use my car over the last 12 months, but I am seriously thinking of doing so over the next 12 months'); 7 per cent of drivers were in this stage.
- *preparation* ('I have already tried to reduce my car use in small ways over the last 12 months and I am planning how to use my car less over the next 12 months'); 22 per cent of drivers were in this stage.
- *action* ('I have tried to reduce my car use over the last 12 months and I am trying to reduce it even more over the next 12 months'); 7 per cent of drivers were in this stage.
- *maintenance* ('I have already reduced my car use as much as I can and I am trying to keep it that way'); 23 per cent of drivers were in this stage.

Participants were also asked to indicate on a 1-4 scale (1 = 'very well' to 4 = 'not well at all'), how well their chosen description fitted them. A large majority (82 per cent) indicated that their chosen description fitted them either 'very well' or 'well'.

Further items were used to measure participants' decisional balance, use of processes of change, and self-efficacy. Decisional balance was measured by 32 items to be rated for importance on a 1-5 scale. Examples were 'If I didn't use the car I would be limited as to where I could go' and 'If I didn't use the car I would be fitter'. Use of processes of change was measured by 36 items to be rated for frequency on a 1-5 scale, where 1 = 'never' and 5 = 'very often'. Examples were 'I walk or cycle for short journeys' and 'I make commitments to reduce my car use'. Self-efficacy in relation to avoiding car use for a short journey was measured by 9 items to be rated for confidence on a 1-5 scale. Examples were 'I would not use the car even if it was raining hard' and 'I would not use the car even if I was feeling tired'.

It was found that those in precontemplation differed from those in the other stages. Compared with those in other stages, the precontemplators used the processes of change least. They attached the greatest importance to the cons of reducing car use and attached the least importance to the pros of reducing car use. In other words, the precontemplators saw the most disadvantages in reducing car use, and the least advantages in doing so. Self-efficacy tended to increase from precontemplation through to maintenance, although once again, the most marked difference was between precontemplators and the rest. There were relatively few significant differences between participants who placed themselves in contemplation, preparation, action or maintenance. This could be partly accounted for by the small number of participants in contemplation and action, resulting in loss of power in the statistical tests. However, the lack of distinguishing characteristics between those beyond the stage of precontemplation raised doubts as to whether these are distinct stages or not. Further work is needed to investigate whether or not there are 5 distinct stages when the model is applied to drivers' willingness to reduce their car use.

The fuel shortages during September 2000 provided the opportunity of following up this earlier questionnaire study in order to find out how drivers at different stages had been affected, and how they coped with any difficulties. The extent and nature of any change in stage was investigated, although it was recognised that there was no comparison group, since everyone was exposed to the fuel shortages unless they were out of the country at the time. Items used to measure processes of change and decisional balance were also included. Any changes between April and

September could not necessarily be attributed to the fuel crisis but could be due to changes in circumstances. The rest of this chapter describes the questionnaire study undertaken during September 2000, the results obtained and conclusions which may be drawn.

Fuel Crisis Study Details

Participants who had responded to the questionnaire study five months earlier (Beatty, 2001) were selected according to their stated willingness to reduce their car use as indicated by their chosen stage of change. On 28th September 2000, questionnaires were sent through Staffordshire University's internal post to 275 employees, 86 of whom had placed themselves in precontemplation, 86 in preparation and 103 in maintenance. The contemplation and action stages were omitted due to the low numbers of potential participants available in these 2 groups. The selection of those in the remaining 3 groups was made on the basis of how well the participants felt they were being described in their chosen stage. All those selected in the preparation and maintenance stages felt they were described 'well' or 'very well'; due to the larger numbers available, all those in the precontemplation stage felt they were described 'very well'.

The questionnaire asked whether travelling arrangements were affected by recent fuel shortages and whether participants had been affected in other ways. In each case, participants who responded 'Yes' were asked to supply details about how they were affected and what measures they took to cope with any difficulties. They were asked whether their opinions about their own use of the car had changed, and whether they had made any lasting changes in travelling behaviour as a result of the fuel shortages. Those who responded 'Yes' were invited to supply details about how their opinions and/or behaviour had changed. It should be kept in mind that only 2 weeks had elapsed between the fuel crisis and the questionnaire study; it would require a follow-up questionnaire to ascertain whether or not reported changes were actually 'lasting'.

Participants were asked to choose the best description of the stage they were now in regarding their willingness to reduce car use. The stage descriptions mirrored those used in the earlier questionnaire study (Beatty, 2001) but they were worded in relation to the fuel crisis e.g. 'I had already reduced my car use as much as possible before the fuel crisis and I shall try to keep it that way' (maintenance). Further items were identical to some of those used on the earlier questionnaire, in order to investigate any changes in the participants' thinking or behaviour in relation to car use. These items were grouped into 2 scales.

The first scale comprised 18 items with a response format of 1-5 to indicate how often the particular statement has applied over the last month, from 1 = 'never' to 5 = 'very often'. These items represented 6 processes of change, measured by 3 items for each process. The processes with their associated items and their reliabilities as measured by Cronbach's alpha[1] were:

Reinforcement management (alpha = 0.75)
- 'I reward myself during the journey if I haven't used the car.'
- 'I treat myself to something if I haven't used the car for a journey.'
- 'I praise myself if I haven't used the car.'

Consciousness raising (alpha = 0.82)
- 'I read books about the effect of private car use on the environment.'
- 'I watch TV programmes about the effect of private car use on the environment.'
- 'I read newspaper or magazine articles about the effect of private car use on the environment.'

Helping relationships (alpha = 0.71)
- 'I offer lifts to those who are making the same journey as I am.'
- 'I share responsibility for dropping off or collecting others with people I know.'
- 'I accept lifts from those who are making the same journey as I am.'

Counterconditioning (alpha = 0.69)
- 'I organise my leisure time so that I avoid unnecessary car journeys.'
- 'I consider whether my journey is really necessary before I use the car.'
- 'I organise my working hours so that I avoid unnecessary car journeys.'

Dramatic relief (alpha = 0.73)
- 'I get cross when other people use their cars for short journeys.'
- 'I get annoyed when I see children being taken to school by car.'
- 'I get cross when I see cars with only one person in them.'

Environmental concern (alpha = 0.66)
- 'I think about the effect that my own car use has on the environment.'
- 'I am aware that my own car use makes the roads congested.'
- 'I get cross when I see fumes from badly maintained cars polluting the air.'

The second scale comprised 12 items with a response format of 1-5 to indicate how important the particular consideration was in deciding whether or not to use the car for a journey, from 1 = 'not at all important' to 5 = 'extremely important'. The items represented 4 aspects of decisional balance i.e. weighing up the pros and cons of using the car for a journey. The 4 aspects with their associated items and their reliabilities as measured by Cronbach's alpha were:

Instrumental/environmental pros (alpha = 0.77)

- 'If I didn't use the car I would not be adding to the pollution.'
- 'If I didn't use the car I would not be adding to the congestion.'
- 'If I didn't use the car I might be supporting public transport.'

Affective (psychological) pros (alpha = 0.75)

- 'If I didn't use the car my family would feel proud of me.'
- 'If I didn't use the car my friends would feel proud of me.'
- 'If I didn't use the car I would feel proud of myself.'

Constraints on work/leisure (alpha = 0.83)

- 'If I didn't use the car it would affect the hours I could work.'
- 'If I didn't use the car my time of travelling would be less flexible.'
- 'If I didn't use the car it would affect the way in which I did my work.'

Journey cons (alpha = 0.78)

- 'If I didn't use the car I would be more exposed to cold weather.'
- 'If I didn't use the car I would have to buy warmer/stronger outdoor clothing.'
- 'If I didn't use the car I would be more exposed to wet weather.'

Fuel Crisis Study Results

Response Rates and Demographics

Questionnaires were returned from 165 participants (60 per cent response rate). Of these, 163 were useable. The participants ranged in age from 20 to 64 years (mean = 42.50; SD = 9.53). They drove up to 30,000 miles a year (mean = 9,679; SD = 5,936). The mean annual mileage is comparable with the National Travel Survey figure of 9,340 miles (DETR, 1999). The sample was 56 per cent female. Table 9.1 shows how many questionnaires were sent out and returned by those judged to be in 3 different stages.

Table 9.1 Response rates for useable questionnaires

Stage in April	Number sent out	Number returned	Response rate (%)
Precontemplation	86	45	52
Preparation	85	51	60
Maintenance	102	67	66

It can be seen that the highest response rate was from maintainers and the lowest response rate was from precontemplators; however, the differences in response rate were not statistically significant.

Whether Travelling Arrangements Were Affected

In response to the question 'Were your travelling arrangements affected by recent fuel shortages', 67 participants (41 per cent) circled 'Yes'. The largest number (27) were from the 67 initially in maintenance, 25 were from the preparation group (out of 51) and 15 were from the precontemplation group (out of 45). Hence the precontemplators were a little under-represented, whereas the preparers were a little over-represented amongst those who responded that their travelling arrangements had been affected, although these differences were not statistically significant.

The main themes emerging from the responses concerned cancelled or altered plans, using alternative forms of transport and finding alternatives to having to travel. A picture emerged of work journeys having priority over all other ones, especially over leisure trips, and of potential journeys being assessed for necessity. Mention was made of coping measures to obtain or conserve fuel.

There were 60 references from 44 participants to specific arrangements for meetings, visits, trips or journeys being cancelled, altered, not attended, postponed or rescheduled, or general statements about travelling less, fuel being conserved for emergency use, and non-essential journeys being cut out.

Leisure trips, weekends away, day trips and social activities were abandoned. For instance:

> Leisure activities … stopped until the fuel crisis did.

> Cancelled a planned weekend in the Lake District.

I was on leave and was unable to take 'day trips' as previously arranged.

To a lesser extent, changes were also made to arrangements at work, by cancelling or postponing meetings, or adapting work travel to reduce mileage. For instance:

Had to send apologies or change dates - this made more work in the end.

Remained on one site, and did not attend meetings. Travelled to nearest site to work.

The importance of cutting out all but essential car journeys and conserving fuel for emergencies, priority journeys or for significant events was highlighted by the following:

Had to cancel other activities apart from necessity.

Reduced use of vehicles during the evening to preserve fuel for travelling to/from work.

Daughter's wedding on the Saturday. All family and guests had to conserve fuel for occasion.

There were 31 references from 25 participants to using other modes of travel, or of sharing or borrowing cars. Walking and using the train were mentioned most frequently; cycling, bus, minibus and taxi were also mentioned. Sometimes an alternative to the car was used every day rather than just sometimes. For instance:

I walked if journey was 2/3 miles and I had the time.

Could not use my car, took bus instead, relied on lifts.

There were 7 participants who used alternatives to making the journey at all, by more use of the telephone, working from home, using video-conferencing and staying overnight. For instance:

Talked on telephone to answer urgent questions only.

Made sure I had no Friday meetings before working at home on that day.

Used video-conferencing for some *(meetings)* but some were outside of the organisation and therefore had to be rearranged …

Three participants specifically mentioned measures to acquire sufficient fuel, by making detours, filling up whenever possible, and siphoning petrol from car to car. Side-effects on travelling arrangements, mentioned by one or two participants included busier trains with standing room only, longer journey times, and having to cover the responsibilities of colleagues who did not have fuel.

Whether Participants Were Affected in Other Ways

In response to the question 'Were you affected in other ways by recent fuel shortages', 45 participants (28 per cent) circled 'Yes'. The largest number (24) were from the 67 initially in maintenance, 15 were from the preparation group (out of 51) and 6 were from the precontemplation group (out of 45). Hence the precontemplators were a little under-represented, whereas the maintainers were a little over-represented amongst those who responded that their travelling arrangements had been affected, although these differences were not statistically significant.

Several responses duplicated or overlapped with the responses to the previous question about travelling arrangements. However, some different themes emerged. The most frequently mentioned ones concerned shortages of supplies, the need for information gathering and advance planning, and worries or anxieties about the situation.

There were 19 references from 17 participants to the shortages of supplies, mainly bread and milk from the shops, but also to eating places being closed and oil for central heating not being delivered. Participants coped in different ways:

> Food and provisions not available at shops. *(Coping measures?)* Family members obtained on our behalf.
>
> Bought dried milk powder and made own bread.
>
> We have oil fired central heating and couldn't get oil delivered for about 2 weeks after the crisis. *(Coping measures?)* Big jumpers.

Planning and information gathering were mentioned by 9 participants. Specific strategies included keeping abreast of the news, enquiring about Lead Replacement Petrol (LRP), making contingency plans and having enough fuel before the crisis started. For instance:

> Check if my car could take LRP petrol, visited the AA website, listened out to TV + radio news programmes.

Organised an international conference that coincided with the crisis. Had back up plans if speakers or equipment didn't materialise.

Five participants referred to being worried or anxious about the situation, either because they themselves might run out of fuel or because they might be affected by others running out, especially when a particular event had been planned. For instance:

Anxiety that crisis would continue and I might run out of fuel. Anxiety that after school kid's care may not be able to pick up my children from school.

Apprehension of my fuel tank running out before strike was called off - particularly as I'm 7 months pregnant.

Our silver wedding party was arranged for the Friday, so spent a long time worrying friends wouldn't come ... friends conserved their fuel so they came anyway.

There were a variety of other ways in which participants were affected. These including experiencing the beneficial effects of not using the car, restrictions on their contact with family members and effects on student enrolment.

Changes of Opinion

There were 32 participants (20 per cent) who responded 'Yes' to whether they had changed their opinions about their own use of the car as a result of recent fuel shortages. The largest number (16) were from the 51 initially in preparation, 12 were from the maintenance group (out of 67) and 4 were from the precontemplation group (out of 45). It can be seen from these figures that the preparers were over-represented in responding that they had changed their opinions about their own car use. A 3 x 2 χ^2 test[2] indicated that the association between stage of change and having changed opinion was significant ($\chi^2 = 7.57$, df = 2, p = 0.02, N = 161). However, this result should be interpreted with caution since, in 5 cases, the details provided described a behavioural change rather than a change of opinion.

Almost all the comments were about feeling the need to think more carefully about car journeys, distinguishing between necessary and unnecessary journeys, and realising the extent of their own and other people's car dependency. The following quotes are typical of a number of responses:

I think more than ever I will think before using the car for unnecessary trips.

> Resolved to look into all possible public transport alternatives for journeys.
>
> We are so dependent on our cars - it was reflected in the selfish way people argued over fuel and hoarded stocks - when others had none.

The extent of their reliance on the car was brought home to some participants, as were the short-comings of public transport. Others realised that they could use the car less if they planned more carefully and that they could affect their fuel consumption by the way they drove. For example:

> Living out in the countryside, I became very aware of how much I need my car and cannot be without it.
>
> Has revealed how pathetic public transport is.
>
> I found that I could manage without my car in certain circumstances which I wouldn't have considered otherwise.
>
> ... have realised, with careful driving and lower speeds, fuel consumption improves.

The fuel shortages therefore forced a number of participants to re-evaluate their use of the car and their travel mode choices. It provided the opportunity to consider and make use of alternatives and for some, those alternatives were found to be satisfactory. However others found that the alternatives were unsatisfactory, merely serving to highlight the extent of their reliance on the car. For a majority of those who indicated that their opinions had changed, it was by way of questioning their assumption that a journey had to be made at all, and made by car.

Lasting Changes in Travelling Behaviour

There were 24 participants (15 per cent) who responded 'Yes' to whether they had made changes in their travelling behaviour which they expected to last beyond the fuel shortages. Once again the largest number (12) were from the 51 initially in preparation, 6 were from the maintenance group (out of 67) and 6 were from the precontemplation group (out of 45). As with those who changed their opinions about car use, it can be seen that the preparers were over-represented in those who responded that they had made lasting changes in their travelling behaviour. The association between stage of change and lasting change in travelling behaviour was not significant ($\chi^2 = 4.86$, df = 2, p = 0.09, N = 162), but came close to the significance level of 0.05 and has therefore been reported.

Three different sorts of behavioural change were mentioned by several participants: reducing car use in a variety of ways, driving in such a way as to conserve fuel and reduce stress, and keeping the tank topped up. Behaviour aimed at reducing car use involved walking, cycling, using the bus or train, combining trips, cutting out unnecessary journeys, and staying overnight. Measures aimed at conserving fuel involved driving more slowly, swapping the second car for a motorbike, and leaving more time for the journey. Some examples of different behavioural changes within one response were:

> Walk whenever possible. Do not make any unnecessary journeys. Always keep tank above half full.

> Keep on driving at lower speeds / accelerating more slowly and ride or walk more often. Try to use public transport more - not easy - especially the trains.

> Will travel by train occasionally (but costs more than car!) for work. Keep fuel topped up.

The majority of participants who reported a lasting change in behaviour described at least one way in which the car could be used less; several others described choosing to drive in a way which conserved fuel. At the same time, the more defensive reaction of keeping an adequate supply of fuel in the tank was also mentioned.

What was the Pattern of Stage Movement?

Questionnaires were sent to participants who had selected one of three stages (precontemplation, preparation or maintenance) when they responded to the earlier questionnaire study (Beatty, 2001). In the present study they were asked to choose the stage description which best fitted them now, in relation to the fuel crisis, out of precontemplation, contemplation, preparation, action and maintenance. There were therefore 15 possible combinations of stage, since each participant could select the same stage as before, or choose any of the remaining 4 stages from 3 possible starting points. Four participants did not select any of the stages. Table 9.2 shows the number of participants making each possible movement.

It can be seen from Table 9.2 that all possible stage movements were represented. Comparing those who had started in the 3 different stages with regard to whether they had changed stage or remained the same, it can be seen that the greatest movement occurred from preparation, where 38/51

(75 per cent) had changed stage. The smallest movement occurred from precontemplation, where 12/42 (29 per cent) had changed stage. Of those who started in maintenance, 28/66 (42 per cent) had changed stage.

In order to test whether the degree of stage movement was statistically significant, a Kruskal-Wallis ANOVA was performed. The degree of stage movement made by each participant who started in either precontemplation or maintenance was calculated as a decimal of the maximum movement possible. This meant that the 2 participants who changed from precontemplation to maintenance were coded as 1 (all possible movement), the 2 participants who changed from precontemplation to action were coded as 0.75 (3/4 of movement possible), the 5 participants who changed from precontemplation to preparation were coded as 0.5 (1/2 of movement possible), the 3 participants who changed from precontemplation to contemplation were coded as 0.25 (1/4 of movement possible) and the 30 participants who remained in precontemplation were coded as 0 (no movement at all).

Table 9.2 Participants' stage movement (n = 159)

Stage chosen in September 2000 (fuel crisis study)	No.	Stage chosen in April 2000 (pre fuel crisis study) Precontemplation n = 42		Preparation n = 51		Maintenance n = 66	
		No.	%	No.	%	No.	%
Precontemplation	50	30	72	10	20	10	15
Contemplation	13	3	7	4	8	6	9
Preparation	26	5	12	13	25	8	12
Action	15	2	5	9	18	4	6
Maintenance	55	2	5	15	29	38	58

This system created difficulties for those in preparation for 2 reasons. Movement could be in either direction but only the degree of movement could be measured. Furthermore, the maximum possible movement from preparation was 2 stages, compared with a possible 4 stages starting from precontemplation or maintenance. In order not to inflate the degree of movement by those who started in preparation, a more conservative estimate of stage movement was used. The change from preparation to contemplation or action (i.e. one stage moved) was coded as 0.25; movement from preparation to precontemplation or maintenance (i.e. 2 stages moved) was coded as 0.5. This created a ceiling effect for those moving from preparation since the maximum score possible was 0.5 compared with 1 for those moving from the other groups.

There was a significant difference in the median degree of movement for those starting in the 3 different stages ($\chi^2 = 10.39$, df = 2, p = 0.006, N = 159). Mann Whitney U tests were used to investigate where the differences between stages lay. Three tests were performed, contrasting precontemplation with preparation, precontemplation with maintenance, and preparation with maintenance. A Bonferroni adjustment of 0.05/3 (0.017) was made to the alpha level, since three contrasts were being performed. The preparers changed stage significantly more than the precontemplators did (U = 650, z = -3.50, $p < 0.001$, N = 93, two-tailed test), despite the fact that the measurement of degree of change from preparation tended to underplay the amount, compared with the other stages. Neither of the other two contrasts was significant.

Table 9.2 showed that, out of the original group of preparers, 24 participants selected later stages (action or maintenance), but there were 14 participants who selected earlier stages (contemplation or precontemplation). Reasons underlying these changes in both directions deserve further investigation, particularly with regard to whether they were associated with an actual decrease or increase in mileage.

Do Those With Different Patterns of Stage Movement Differ in Other Ways?

Table 9.2 showed that all 15 possible patterns of movement were represented. In order to arrive at a smaller number of groups for further analysis, several smaller categories were collapsed together to produce a total of 7 groups. Table 9.3 shows the composition of these groups and the number of participants in each.

Possible differences were explored between these 7 groups in their use of the 6 processes of change and the 4 aspects of decisional balance. Ten mixed ANOVAs were conducted, with group as a between-subjects measure (7 levels) and time of year as a within-subjects measure (2 levels).

In the case of 3 of the processes of change, reinforcement management, dramatic relief and environmental concern, there was an interaction between time of year and group which, although not significant, looked worthy of further investigation by looking for simple effects as recommended by Clark-Carter (1997). The results for the interactions were:

- Reinforcement management: $F_{(1,6)} = 1.41$, $p = 0.21$, $MS_{interaction} = 0.34$.
- Dramatic relief: $F_{(1,6)} = 1.63$, $p = 0.14$, $MS_{interaction} = 0.46$.
- Environmental concern: $F_{(1,6)} = 1.32$, $p = 0.25$, $MS_{interaction} = 0.32$.

Table 9.3 Composition of groups according to pattern of movement

Gp.	Stage in April 2000 (pre fuel crisis study)	Stage in Sept. 2000 (fuel crisis study)	Possible interpretation of change	No. (%)
1	Precontemplation	Precontemplation	No movement	30 (19)
2	Precontemplation	Any other stage	Move to later stage	12 (8)
3	Preparation	Precontemplation or contemplation	Move to earlier stage	14 (9)
4	Preparation	Preparation	No movement	13 (8)
5	Preparation	Action or maintenance	Move to later stage	24 (15)
6	Maintenance	Any other stage	Move to earlier stage.	28 (18)
7	Maintenance	Maintenance	No movement	38 (24)

Accordingly, the effect of time on each of the 7 groups was investigated using 7 paired t-tests, one for each of the 7 groups. It should be noted that by making 7 paired t-tests after each ANOVA, the chance of a type 1 error has increased. However, a Bonferroni adjustment in the alpha level to 0.05/7 (i.e. 0.007) might be regarded as overly conservative and has therefore not been used. However, any significant differences should be interpreted with caution.

The participants in group 2 (precontemplators who had changed to a later stage) used reinforcement management significantly more in September 2000 than in April 2000 ($t_{(11)} = -2.41$, $p = 0.02$, one-tailed test).

The participants in group 7 (maintainers who stayed in maintenance) used dramatic relief significantly more in September 2000 than in April 2000 ($t_{(36)} = -3.31$, $p = 0.001$, one-tailed test).

The participants in group 4 (preparers who stayed in preparation) used environmental concern significantly more in September 2000 than in April 2000 ($t_{(11)} = -2.62$, $p = 0.01$, one-tailed test). Also, participants in group 6 (maintainers who moved to an earlier stage) used this process significantly more in September 2000 than in April 2000 ($t_{(27)} = -1.82$, $p = 0.04$, one-tailed test). However, bearing in mind that 7 t-tests have been performed, this last effect may not be reliable since the p value is only slightly below 0.05.

Conclusions

The questionnaire circulated to selected employees of Staffordshire University during September 2000 provided an unusual opportunity to assess the impact of the fuel shortages on drivers' behaviour and opinions. It also provided the opportunity to compare participants' responses with those obtained 5 months earlier (Beatty, 2001) and to investigate differences between those in different stages of change according to the transtheoretical model (Prochaska and DiClemente, 1982).

Before considering the implications of the findings of this study it is worth considering its methodological weaknesses. One limitation is that only employees working for one organisation, a University, were studied, although within that organisation there were people doing a variety of types of work e.g. academic, administrative, clerical, managerial, manual, supervisory and technical. The sample did not include the retired, the unemployed or those whose work was entirely home-based, all of whom might have been affected to different extents by the fuel shortages.

Compared with the stages reported in the pre-fuel crisis data (Beatty, 2001) there was a lot of apparent change over a relatively short period of time, and not necessarily in an expected direction. One possible interpretation of this situation is that these apparent changes in car use are actually happening, and that people move in both directions regarding willingness to reduce their car use. In support of this interpretation is that the model has been conceptualised as a spiral (Prochaska et al., 1994a) or a revolving door (Prochaska and DiClemente, 1982) and does allow people to move back through earlier stages several times. In attempting to give up an addictive behaviour, it would be common to spiral though the model several times before reaching termination, when there is no temptation to return to the problem behaviour. The fuel crisis itself might have made some drivers realise the lack of satisfactory alternatives to the car and consequently their dependency on it, thus encouraging them to select an earlier stage compared with their response in April when there was no threat to their car use.

The erratic nature of the stage movements also raises questions about the operationalisation of the stages when the model is applied to car use, and whether the questionnaire contained a valid measure of the stages. More needs to be known about why the participants appeared to be changing stage and to what extent they felt that their chosen stage description fitted them well. Although the pre-fuel crisis study had asked participants how well their chosen stage fitted them, those selected for the fuel crisis study were not asked about this again. In retrospect, it would have been useful to do so and to conduct follow up interviews with those who did not feel they were well described in order to investigate these issues. Follow up interviews could also provide information about whether more lasting changes of opinion or travelling behaviour were still present several months later.

A particular difficulty in operationisation might relate to the maintenance stage, since reducing one's car use as much as possible is going to mean different things to different people. In a study of dietary fat reduction, Brug et al. (1997) found that a substantial number of participants in maintenance actually consumed much more fat than was recommended. Therefore some maintainers should be regarded as unaware precontemplators. However, the April study (Beatty, 2001) found that drivers classified as being maintenance had a significantly lower annual mileage than drivers in precontemplation, contemplation or preparation. This suggested that the choice of stage did bear some relationship to miles driven, and that those in later stages drove less than those in earlier stages.

A crucial issue is to consider the natural differences in circumstances between April and September, irrespective of the fuel crisis. As acknowledged earlier, because there was no control group, it was not possible to compare those who experienced the fuel crisis with those who did not. Only the few who were out of the country at the time escaped its effects completely. Different demands due to the different times of year might have been expected to account for some of the changes e.g. University employees with responsibilities for students would have different demands on their time and different degrees of flexibility depending on time of year.

An additional consideration is that people's circumstances vary greatly, as would the extent to which they were prepared, accidentally or otherwise, for the fuel shortages. Someone with a full tank of fuel, access to satisfactory public transport and only short journeys to make would be in a very different position to someone with an almost empty tank, living in a rural area and with pressing engagements. The level of anxiety might well relate to the ease with which alternatives to the car could be found for important events already planned. Participants were invited to explain how they were affected by the fuel crisis. There were no striking differences in the responses of participants originally judged to be in 3 different stages. In any case, larger numbers would be needed in order to make such comparisons meaningful.

More changes were to have been expected in the use of processes of change and decisional balance by different groups over time. Low numbers in many of the groups made analysis difficult and compromises had to be made by combining smaller groups together for analysis. Clearly, further work is needed to judge whether the items chosen for decisional balance and for the processes of change are appropriate. Indeed, it may be that not all the processes can be applied to reducing car use. Prochaska and Velicer (1997) pointed out that, depending which behaviour was being studied, some studies have found that fewer processes are used, and sometimes there has been evidence for additional processes.

This study suggested that many people, at least in the short-term, have both the capacity and capability to reduce their car use when they are forced to do so. This supports Goodwin's (1995) analysis of car dependency as involving considerable scope for reduction by focusing on the large percentage of trips which do not have to be made by car. Having identified the trips which could be made by alternative means, or not made at all, drivers need to be encouraged to continue in the strategies they used during the fuel crisis.

The travel blending approach (Ampt, 1999) used with Australian households has been found to be a useful means of encouraging people to assess their travelling habits and make manageable changes which fit into their patterns of work and leisure. The enforced assessment of car use which occurred during the fuel crisis, provided the opportunity for households to see what longer-term changes they might make, although only 15 per cent of participants thought that there would be lasting changes in their travelling behaviour. As pointed out earlier, a period of time considerably longer than 2 weeks would be needed to find out whether lasting change had indeed been effected. Strategies used to cope with the crisis in the short-term might not have been so attractive when applied over a longer time period.

Regarding drivers as being in stages in their willingness to reduce car use offers suggestions as to how different groups of drivers might be targeted. Preparers would appear to be a worthwhile group to target in that they were more receptive and were most likely to change stage - either forwards or backwards. They were also more likely to say that they had changed their travelling behaviour in lasting ways and changed their opinions during the fuel crisis. In the sense that preparers were defined as those who had already made small reductions in their car use and were planning how to make more, the fuel crisis provided the incentive to put that planning into practice. The question remains as to why so many preparers moved to an earlier stage in the model.

One aim of practitioners wanting to encourage modal shift away from the car would be to target the largest group, the precontemplators, and try to move them to later stages of the model. Analysis showed that those precontemplators who moved to later stages during the fuel crisis increased their use of reinforcement management. Reinforcement management is a technique whereby drivers might give themselves rewards for not using the car, either in praise or in more tangible forms. Encouraging the use of reinforcement management may help drivers in precontemplation to reduce their car use.

However, precontemplators were the least likely to change stage. They also reported least effects on their travel and other arrangements during the fuel crisis. Since, by definition, precontemplators had neither tried to reduce, nor were thinking of reducing their car use, it was logical to find that they were the least likely to change stage. It is more difficult to understand why precontemplators reported less effect on their travelling arrangements, since the pre-fuel crisis study (Beatty, 2001) found that precontemplators had a significantly higher annual mileage than maintainers. One possibility is that they were least willing to accept having to reduce their car use, and took more drastic steps to continue with their plans.

A large proportion of those drivers in maintenance reverted back to earlier stages of the model during the fuel crisis. Reasons for this deserve exploration. It is not known whether the fuel crisis itself triggered a change, or whether a change had occurred at some earlier point. Those maintainers who did not relapse were significantly more likely to experience dramatic relief, that is they were more likely to respond emotionally to what they saw as other people's inappropriate use of the car. Interventions might play upon this to encourage maintenance of low car use.

It is unclear how drivers in different stages are differentially concerned with the environment. Both preparers who stayed in preparation, and maintainers who moved to an earlier stage, reported more environmental concern during the fuel crisis. Why there should be this difference in these groups but not in the others requires further investigation. In the pre-fuel crisis study (Beatty, 2001) it was found that environmental concern was reported more than any other process by drivers at all stages, suggesting scope for messages encouraging drivers to reduce their car use for environmental reasons.

The fuel shortages of September 2000 highlighted for many people their own and their country's reliance on the car. The shortcomings of public transport were brought home to some of those who would not normally use it, so it is clear that public transport needs to be more reliable if it is to attract the occasional user on a more regular basis. Whether those of us who are car users revert to our previous travelling behaviours and forget the warning signs about the car-dependent lifestyles we and future generations are becoming enmeshed in is the choice we all have. Government, local authorities, public transport operators and employers can also learn their lessons from the short period of time when the country found out just how much it depended on having an adequate supply of fuel.

Notes

1 Cronbach's alpha is a measure of a scale's internal consistency i.e. the degree to which the scale's items hang together (Pallant, 2001). The alpha coefficient is measured on a linear scale running from 1 (totally reliable) to 0 (totally unreliable). A figure above 0.7 is generally thought to be acceptable. However, Cronbach's alpha is sensitive to the number of items in a scale, with shorter scales tending to produce lower values of alpha.

2 Readers wishing to know more about the χ^2 test, and other statistical tests mentioned in this chapter, are recommended to refer to Clark-Carter (1997).

References

Ampt, E. (1999). Grass routes: From travel blending to living neighbourhoods. *Traffic Engineering and Control*, 40, 475-478.

Bandura, A. (1997). *Self-efficacy. The exercise of control.* New York, W.H. Freeman and Company.

Beatty, S.F. (2001). *Psychological factors and stages of change in drivers' willingness to reduce their car use.* Unpublished doctoral dissertation, Staffordshire University, England.

Brug, J., Hospers H.J. and Kok, G. (1997). Differences in psychosocial factors and fat consumption between stages of change for fat reduction. *Psychology and Health*, 12, 719-727.

Cardinal, B.J. and Sachs, M.L. (1996). Effects of mail-mediated, stage-matched exercise behavior change strategies on female adults' leisure-time exercise behavior. *Journal of Sports Medicine and Physical Fitness*, 36, 100-107.

Clark-Carter, D. (1997). *Doing quantitative psychological research.* Hove, Psychology Press Ltd.

Clarke, P. and Eves, F. (1997). Applying the transtheoretical model to the study of exercise on prescription. *Journal of Health Psychology*, 2, 195-207.

DETR (1998). *A New Deal for Transport: Better for Everyone.* London, TSO.

DETR (1999). *National Travel Survey: 1996/98 update.* London, DETR.

DETR (2000). News Release 640: *Transport Statistics Great Britain 2000 - 26th Edition.* Available at: http://www.press.detr.gov.uk/0010/0640.htm [accessed 29/11/00].

Fava, J.L., Velicer, W.F. and Prochaska, J.O. (1995). Applying the transtheoretical model to a representative sample of smokers. *Addictive Behaviors*, 20, 189-203.

Fergusson, M., Davis, A. and Skinner, I. (1999). *Delivering changes in travel behaviour.* London, Institute for European Environmental Policy.

Goodwin, P. (Ed.) (1995). *Car Dependence.* RAC Foundation for Motoring and the Environment.

Holz, S.B. (1996). *Understanding and using the stages of change.* Available at: http://www.opc.on.ca/ptcc/webdocs/stages/changes.html [accessed 25/11/99].

Lex Report on Motoring (1998). *Driving for the future.* Lex Service PLC, with research by MORI.

Mutrie, N. Blamey, A. and Whitelaw, A. (1999). *A randomised controlled trial of a cognitive behavioural intervention aimed at increasing active commuting in a workplace setting.* Report to Chief Scientist Office, Scottish Executive.

Pallant, J. (2001). *SPSS survival manual.* Buckingham, Open University Press.

Perz, C.A., DiClemente, C.C. and Carbonari, J.P. (1996). Doing the right thing at the right time? The interaction of stages and processes of change in successful smoking cessation. *Health Psychology*, 15, 462-468.

Povey, R., Conner, M., Sparks, P., James, R. and Shepherd, R. (1999). A critical examination of the application of the transtheoretical model's stages of change to dietary behaviours. *Health Education Research*, 14, 641-651.

Prochaska, J.O. and DiClemente, C.C. (1982). Transtheoretical therapy: toward a more integrative model of change. *Psychotherapy: Theory, Research and Practice*, 19, 276-289.

Prochaska, J.O. and DiClemente, C.C. (1983). Stages and processes of self-change of smoking: toward an integrative model of change. *Journal of Consulting and Clinical Psychology*, 51, 390-395.

Prochaska, J.O., Norcross, J.C. and DiClemente, C.C. (1994a). *Changing for Good.* New York, Avon Books.

Prochaska, J.O. and Velicer, W.F. (1997). The transtheoretical model of health behavior change. *American Journal of Health Promotion*, 12, 38-48.

Prochaska, J.O., Velicer, W.F., Guadagnoli, E., Rossi, J.S. and DiClemente, C.C. (1991). Patterns of change: dynamic typology applied to smoking cessation. *Multivariate Behavioral Research*, 26, 83-107.

Prochaska, J.O., Velicer, W.F., Rossi, J.S., Goldstein, M.G., Marcus, B.H., Rakowski, W., Fiore, C., Harlow, L.L., Redding, C.A., Rosenbloom, D. and Rossi, S.R. (1994b). Stages of change and decisional balance for 12 problem behaviors. *Health Psychology*, 13, 39-46.

10 Effective Fuel Management in Road Transport Fleets

TONY WHITEING, MICHAEL COYLE AND COLIN BAMFORD

Introduction

As the September 2000 fuel crisis in the UK started to bite, freight transport and logistics operators faced something of a dilemma. Many appeared to have considerable sympathy with the protesters picketing the fuel distribution depots. Indeed, many of them had participated in protests over the high price of fuel. Operators had joined protest convoys in several major cities, including London and Edinburgh. Many supported the Road Haulage Association's 'Fair Play on Fuel' campaign (RHA, 2000), which had been running for some time prior to the crisis. There was a widely held view that the UK freight transport industry was a victim of unfair competition. Fuel taxes elsewhere in Europe are often significantly lower than in the UK, even where governments had signed up to the Rio and Kyoto accords. Other tax differences such as relatively high vehicle excise duties have tended to add to international operating cost differentials, which are high even when the absence of vignette systems or any significant road tolling in the UK are taken into account (RHA, 2000). Irish and French hauliers were perceived to have distinct cost advantages over their UK counterparts, and were seen to be competing very effectively in the UK. There were many tales of UK hauliers refuelling their vehicles whilst abroad in an attempt to remain competitive, but by no means all operators could take advantage of this. A few operators had moved their operating centres and vehicle registrations onto continental Europe (so called 'flagging-out') in an attempt to reduce costs.

Fuel costs form a relatively high proportion of the total costs of operating commercial vehicles in the UK. Table 10.1 shows estimates of fuel costs as a proportion of total operating costs for two popular vehicle types and for a range of annual distances travelled.

Table 10.1 Fuel costs as a proportion of commercial vehicle operating costs, 1996-2000

	1996		1997		1998		1999		2000	
	17 tonne rigid	32.5 tonne articulated	17 tonne rigid	32.5 tonne articulated	17 tonne rigid	32.5 tonne articulated	17 tonne rigid	32.5 tonne articulated	17 tonne rigid	32.5 tonne articulated
Standing costs [a] (£/year)	33654	40015	34668	42784	36069	43411	33853	41676	34445	42657
Running costs [b] (£/mile)	0.3771	0.4247	0.3586	0.4458	0.3642	0.453	0.3655	0.4538	0.4138	0.5162
of which fuel costs (£/mile)	0.2198	0.2585	0.2181	0.2822	0.2236	0.2893	0.2263	0.2928	0.2738	0.3544
Fuel costs as % of operating cost [c]										
60000 miles p.a.	23.4	23.7	23.3	24.4	23.2	24.6	24.3	25.5	27.7	28.9
80000 miles p.a.	27.6	27.9	27.5	28.8	27.4	29.1	28.7	30.0	32.4	33.8
100000 miles p.a.	30.8	31.3	30.9	32.3	30.8	32.6	32.1	33.6	36.1	37.6
120000 miles p.a.	33.4	34.1	33.7	35.2	33.6	35.5	34.9	36.5	39.1	40.7

[a] Standing costs are wages, National Insurance, vehicle insurance, establishment costs, licencesand depreciation
[b] Running costs cover fuel, oil, tyres and maintenance
[c] Operating cost = standing costs plus running costs as defined above

Source: Motor Transport Cost Tables, February 1996, February 1997, February 1998, February 1999 and February 2000

Between 1996 and 1998 fuel prices increased, reflecting the government's 'fuel price escalator', which increased fuel duties by 6 per cent annually over and above the general rate of inflation. This has led to increased operating costs, but the cost of fuel as a percentage of total costs increased only slowly because standing costs were rising over the same period. Between 1998 and 1999 there was a reduction in standing costs as defined in the 'Motor Transport' operating cost tables, due to lower list prices for new vehicles. Fuel costs as a percentage of total costs increased as a result. By far the most notable feature of Table 10.1 is the increase in the cost of fuel as a percentage of total costs between 1999 and 2000. Over the five-year period from February 1996 to February 2000, the cost of fuel for a 32.5 tonne articulated lorry travelling 80,000 miles per annum increased from 28 per cent of total operating costs to nearly 34 per cent. For fleets of high mileage vehicles, fuel bills constituted more than 40 per cent of total operating costs. For operators of large fleets, able to negotiate significant discounts on the list price of new vehicles, fuel costs would form an even greater percentage of total operating costs.

Operators concerned about staff retention were hardly likely to instruct their drivers to attempt to cross the protesters' picket lines. In many parts of the UK, the problems caused by rising fuel costs were being compounded by driver shortages, leading to increased labour costs, recruiting costs and extensive use of agency labour. In such circumstances there is little wonder that the freight transport industry was becoming increasingly vociferous in its claims that unfair competition was eroding the already slender profit margins. Financial results from 153 leading UK haulage firms reveal a fall in pre-tax profit margins from 3.3 per cent in 1996/97 to just 2.8 per cent in 1998/99 (Commercial Motor, 2000).

The opposite side of the coin is of course that the fuel crisis severely jeopardised the freight transport industry's ability to satisfy the demands of its customers. These demands have become increasingly stringent in recent years. Many manufacturers have adopted Just-in-Time production systems, and the major retailers have introduced Quick Response logistics systems. The essence of such systems is that they attempt to minimise stockholding by means of frequent replenishment from suppliers. Highly reliable freight transport services are called for and time windows on delivery are often very narrow (Fowkes et al., 2001). Failure to deliver on time may lead to empty supermarket shelves or to costly industrial processes grinding to a halt. In such circumstances, many manufacturers and retailers have deliberately opted out of transport operations, replacing own account fleets with the use of third party contract logistics services (Bamford, 1999). Hence they have put great pressure on the third party sector to maintain high levels of service whilst distancing themselves from the immediate problems of transport operations, such as fuel shortages. Fowkes et al. (2001) identify various other reasons why the freight transport industry values highly reliable and predictable operating conditions. Examples include the need to maximise labour productivity and the need to arrive at ports in good time for ferry sailings.

Against this background, this chapter presents the findings of a survey of 240 UK operators' responses to the fuel crisis, conducted in May 2001. It also reviews the range of fuel saving interventions in use or under consideration by these operators, and discusses the extent to which these will help to keep costs under control.

A Survey of the Fuel Crisis Impact on Freight Transport Operators

To understand the impact of the fuel crisis on the freight transport industry and to discover its responses to the resulting fuel shortages, a survey of 240 operators was conducted during May 2001. Whilst the immediate problems caused by the crisis were of major interest, the timing of the survey allowed investigation into more measured responses once the crisis had abated. The severity of the crisis ensured that respondents could readily recall its effects on their operations and the measures that were put in place.

A short (two-page) questionnaire was sent to a range of operators of commercial vehicles, covering both the own account and third party sectors. Companies were sampled from the ABC Freight Guide (2000), a commercial directory of the freight transport industry. This was supplemented by companies on three University of Huddersfield databases. These related to companies employing Huddersfield graduates, companies who had sent delegates to a range of conferences and short courses, and companies employing students during their industrial placement year. Companies sampled from the ABC Freight Guide were sent the questionnaire by post. Most companies selected from the University databases also received the questionnaire by this means, though some were approached via email instead.

As shown in Table 10.2, there was a 24 per cent response rate to the survey, providing 58 usable responses. The percentage response by email was higher than that by post, reflecting the fact that the respondents knew the researchers concerned.

Initial questions sought information about fleet size, the nature of the company's commercial vehicle operations and the normal methods used to acquire fuel. Table 10.3 shows the division of respondents between own account and third party sectors, and the business sectors represented.

Table 10.2 Survey response rates

		of which by post	of which by email
Questionnaires despatched	240	219	21
Returned completed	58	47	11
Recipient did not operate commercial vehicles	4	4	0
Questionnaire not delivered	1	1	0
Response rate (per cent) (Returned completed as percentage of those delivered)	24	21	52

Table 10.3 Survey respondents – nature of operation

	Own account	Third party
Total	22	37
of which		
Manufacturing logistics	7	9
Food supply chain	7	15
Retail (non-food)	5	6
Contract logistics	n/a	17
General haulage	n/a	18
Intermodal operations	n/a	6
Other	4	4

n/a – not applicable to 'own account' sector

There were rather more third party sector respondents than own account respondents, roughly in line with each sector's share of the road freight transport market (Bamford, 1999). Own account operators were predominantly from the manufacturing, food and retail industries. Third party companies were involved in a wide range of business. Some concentrated on particular market segments such as the food supply chain, but many undertook general haulage work, or had longer-term logistics contracts with a range of industrial or retail clients. One respondent operated in both the own account and third party sectors.

Table 10.4 Survey respondents – distribution of fleet sizes

Fleet size	Own account	Third party	Total
1-5 vehicles	2	1	3
6-20	5	4	9
21-50*	5*	9*	13*
51-100	4	9	13
101-200	1	6	7
201-500	0	7	7
501 +	4	1	5
Not known	1	0	1

* one company in this size category operates in both own account and third party sectors

The distribution of fleet sizes by sector is shown in Table 10.4. The sample is not claimed as a statistically accurate representation of the true industry population. In particular, the sampling process was not amenable to the capture of small businesses. Hence most fleets had more than 5 vehicles. Indeed, some of largest and most significant UK fleets in both the own account and third party sectors were represented. Altogether, the 58 respondents were responsible for operating over 9,000 vehicles, of which some 5,000 were in the third party sector.

The dominance of larger fleets in the survey is reflected in the fact that most respondents' companies hold stocks of fuel, purchased in bulk from suppliers (Table 10.5). Many also use fuel cards for purchases 'on the road', though relatively few depend on this method alone. A small number maintain accounts at local filling stations, but none use this method exclusively. No respondents reported payment for fuel with cash.

Table 10.5 Survey respondents – methods of fuel purchase

Method of fuel purchase	Number of companies
Bulk purchases only	15
Fuel cards only	7
Account at filling station only	0
Bulk purchase and fuel cards	33
Bulk purchase, fuel cards and account at filling station	1
Fuel cards and account at filling station	2

Remaining questions then addressed the issues relating to the fuel crisis, divided into three time periods. The first of these related to the days directly affected by the fuel blockades. Questions focused on the company's ability to maintain effective operations. The second set of questions related to the strategies adopted in the 60-day period following the fuel blockades, during which the UK government considered its response to the fuel protesters' ultimatum. The third set addressed changes implemented by the operator in the period after the end of the 60-day ultimatum period, up to May 2001. Attention here was focused on strategies to economise on the use of fuel. This aimed to gather information on whether the fuel crisis had spurred the industry into adopting a range of fuel-saving interventions. The response could then be evaluated using available evidence on the fuel savings likely to be gained through such interventions. The remainder of this chapter is structured around these three time periods, before finally arriving at general conclusions.

An outcome of the sampling process was that a number of respondents had previous or ongoing involvement with the University of Huddersfield on a range of fuel efficiency projects. Hence a small amount of analysis was undertaken to investigate whether the behaviour of these nine respondents differed from that of the majority of those who provided information. In particular, it might be hypothesised that the companies involved would be more aware of the benefits of fuel efficiency interventions and therefore have more interventions in place.

The Immediate Impact of the Fuel Crisis

As indicated in Table 10.6 below, almost two-thirds of those surveyed reported that the fuel blockades had impacted on their business operations in one way or another. No one claimed to have built up stocks because they had seen the crisis looming. Only five operators actually ran out of fuel, though some others reported that fuel stocks were very close to zero when the blockades were lifted. Twelve respondents reported that their companies had to turn away work, for instance to conserve stocks for their priority customers. Five reported that work had to be subcontracted because of depleting fuel stocks. Eleven companies imposed speed restrictions in an attempt to conserve fuel.

Respondents were asked to state other impacts of the fuel crisis, and the majority of those affected provided such information which is also shown in Table 10.6. As can be seen, a small number of companies were able to purchase fuel supplies abroad. Several reported that they had to move fuel stocks between their operating centres. Others ensured that vehicles were only despatched when fully laden, presumably in consultation with their customers. Companies running short of fuel naturally tended to concentrate on meeting the needs of their most important customers. A number of respondents were keen to point out that the crisis had increased their costs and had also impacted negatively on the service levels provided to customers. Three companies reported acquiring stocks of duty-free 'red' diesel or gas oil, although only one admitted to using such fuels in their vehicles. Not all impacts were negative. Some operators fortunate to have fuel picked up work that other companies were unable to accept. A small number of own account operators who also use third party hauliers increased their own workload because their hauliers were short of fuel.

Table 10.6 Impacts during the fuel blockades

Impact on operations	No. of respondents affected	% of respondents
No effect on operations	20	34
Ran out of fuel	5	9
Had to decline work	12	21
Had to subcontract work	5	9
Imposed speed restrictions to conserve fuel	11	19
Other impacts stated by operators	29	50
Of which:		
Had to move fuel between sites	6	-
Had to change delivery times	5	-
Had to obtain fuel abroad	4	-
Had to co-operate with other operators to ensure fuel supplies	3	-
Had to amalgamate delivery routes	3	-
Had to send drivers to search for fuel	3	-
Designated as priority need for fuel	17	29
Received priority / emergency supplies	9	16

Nearly a third of survey respondents were designated as priority cases for fuel deliveries. Typically, these were involved in the food supply chain or medical supplies. In the event, only half of these actually received emergency fuel deliveries during the blockade period. Whilst at first sight this may seem surprising, it must be remembered that a large majority of respondents did not in fact run out of fuel. Eight operators not designated as priority cases felt that they should have been. The basis of their claims is not clear, though most of the operators concerned were involved in the food industry.

Strategies Used in the 60-Day Ultimatum Period

The government's response to the demands of the fuel protesters was by no means predictable and therefore the 60-day period following the lifting of the fuel blockades was one of great uncertainty for operators. Hence one might predict that extra fuel would be stocked as a contingency against further supply disruption. This would be particularly true for larger operators, which featured prominently in this survey, because (as shown in Table 10.5) many possess bulk fuel tanks at their depots.

Table 10.7 shows that the vast majority took some action during this period. Significant numbers changed their sources of fuel and/or the methods of payment, but the dominant change was to the amount of fuel held in stock. Over forty respondents reported a change to their fuel stockholding policy. A small number reported reduced stockholdings, but the vast majority increased the level of stock, in some cases very substantially.

Some companies reported that bulk fuel tanks were installed specially so that bulk purchasing could be introduced. This is a decision that would not be taken lightly. Apart from the capital costs of installation, there is a considerable amount of safety legislation that must be complied with.

Table 10.7 Policies towards fuel purchase and stockholding during the 60-day ultimatum period

	No. of respondents	%
No change to policy	8	14
Some change(s) to policy	50	86
of which		
Change to sources of fuel	22	44
Change to fuel stock held	46	92
Change to methods of payment	10	20
Other changes	8	16

Another strategy was to increase the use of fuel cards. Drivers refuelled vehicles in the course of their journeys, in order to conserve stocks in the company's bulk tanks. This might be expected to increase the cost of fuel marginally.

Some changes to suppliers and methods of payment occurred because operators felt that they had been let down or had been given low priority by their existing suppliers.

These somewhat predictable reactions to the uncertainty of the time appear to have cost operators not insignificant amounts of money. Investment in bulk tanks, more purchases 'on the road' and higher inventory financing costs will all have raised the costs of the operators concerned.

Strategies Since The Fuel Crisis

The survey results suggest a rather mixed set of responses following the uneventful passing of the 60-day period. Table 10.8 shows that at the time of the survey (May 2001) 22 operators were still opting to hold fuel stocks higher than the pre-crisis level. Relatively small numbers of operators had decided on other risk avoidance strategies, for example to retain more bulk fuel suppliers or to widen their use of fuel card systems (and to avoid dependence on any one such system). Typically though, fuel stocks had been reduced below the levels of the 60-day period. Around half the respondents appeared to have reverted to the policies in place prior to the crisis.

Table 10.8 Policy changes still in place subsequent to the 60-day period

	No. of respondents
Higher fuel stocks than before crisis	22
Using more or different bulk fuel suppliers	5
More use of fuel cards	3

Fuel Crisis – An Impetus to the use of Fuel Efficiency Interventions?

Survey respondents were asked to provide information on which of a range of potential fuel saving interventions were either in use or under consideration, both before and after the fuel crisis. The results are shown in Table 10.9.

The three most popular interventions were driver training, the use of aerodynamic vehicle styling and the application of computerised routing and scheduling packages. None of these saw significantly increased use subsequent to the fuel crisis, though two companies reported new interest in driver training and two in routing and scheduling packages.

Survey respondents have however increased their use of on-board fuel metering and vehicle tracking systems. Some operators have revised their choice of vehicle manufacturers or the specification of their vehicles. More operators are using or considering the use of vehicles powered by alternative fuels such as compressed natural gas.

Table 10.9 Fuel efficiency interventions in use or under consideration before and after the fuel crisis

Type of fuel efficiency intervention	Before crisis		After crisis	
	No. of operators using	No. of operators considering	No. of operators using	No. of operators considering
Driver training	32	7	33	9
Aerodynamics	32	1	32	1
On-board fuel metering	18	5	22	5
Change of vehicle manufacturers	11	11	15	9
Improved routing and scheduling systems	26	2	27	4
Revised vehicle specifications	12	8	13	9
Use of alternative fuels	2	10	4	13
Vehicle tracking systems	12	11	15	12
Other	2	0	3	1

There is some evidence from the survey that those respondents involved with the University's research into fuel efficiency employ more than the average number of fuel saving interventions. After the fuel crisis, these nine respondents were using or considering an average of 4.4 interventions, compared to an average of 3.9 for the full sample of 58 respondents.

There is little doubt that driver training is one of the most effective ways to reduce fuel consumption. Research undertaken by the University of Huddersfield in conjunction with a major third party logistics operator suggests average fuel savings of around 13 – 14 per cent from a sustained driver development programme (Coyle, 2001). Potential savings will of course vary depending on operating conditions. It is also important to provide regular refresher courses, as the impact of training will fade as time passes.

Evidence on the effect of aerodynamic aids is rather more mixed. Projects undertaken for the Energy Efficiency Demonstration Scheme yielded fuel savings of 16 per cent for articulated vehicles and in excess of 19 per cent for larger rigid vehicles (Department of Energy, 1990, 1992). Work undertaken by Cooper and Tweddle (1994) on the other hand suggested much more modest gains of around 3 per cent for daytime operations and only 1.5 per cent for night trunking operations. The benefits will depend on average vehicle speed. Basic aerodynamic aids yield the major part of the fuel saving. More comprehensive aerodynamic styling produces little extra saving in comparison.

The use of on-board fuel consumption metering equipment appears to be increasing. A case study undertaken by the Energy Efficiency Best Practice Programme reports fuel savings of 8.4 per cent for rigid vehicles and 5.2 per cent for articulated lorries, producing an overall saving of 5.8 per cent for the operator concerned (DETR, 1997). Results from on-board metering can improve fuel consumption in various ways. They can, for example, be fed instantaneously back to the driver via an in-cab display. Results can also be analysed to determine the extent to which the driver is staying within efficient driving parameters. League tables of driver performance can be derived and linked to training schemes and possibly also to incentive payments.

Comparative fuel consumption of different manufacturers' vehicles is a highly sensitive issue, but it is likely that fuel savings in excess of 10 per cent are possible through brand switching, and there is not necessarily any correlation between fuel efficiency and vehicle purchase price.

Sophisticated computerised routing and scheduling packages can lead to very significant improvements in fuel efficiency. Coyle (2000) reports the case of an own account operator in the manufacturing sector, who achieved a 25 per cent reduction in vehicle kilometres operated, equivalent to a reduction of 800,000 kilometres per annum. In many cases, fleet sizes can be reduced, leading to further savings in vehicle standing costs.

Optimum vehicle specifications normally hinge on finding the best combination of engine, gearbox and final drive ratio. Work undertaken by the University of Huddersfield in conjunction with a UK supermarket retailer identified a combination that was 2.2 per cent more fuel efficient than the nearest rival.

Vehicle tracking systems are a relatively new phenomenon and have only recently achieved significant market penetration. They offer the opportunity to identify drivers who deviate from the optimum route. It seems likely that they offer most scope for fuel savings if they incorporate facilities for engine management and reporting. Excessive engine idling speed can be identified, for example. Over-revving, harsh acceleration and severe braking can also be monitored and results used to tailor driver-training programmes.

Alternative vehicle fuels provide well-known environmental benefits. At present their cost effectiveness depends on the availability of grant-aid. There are also significant issues concerned with distances between refuelling and the number of refuelling stations available.

It is evident that whilst some survey respondents make good use of such fuel saving interventions, many could do a lot more. The survey results call for careful interpretation. The use of many such interventions (notably the IT-based interventions such as routing and scheduling packages and vehicle tracking systems) is increasing on trend, so that the link with fuel prices and the fuel crisis is uncertain. It does seem likely however that (at the margin) the high cost of fuel has spurred operators to consider more fuel saving interventions or to bring forward plans for implementation. Operators appear to be monitoring fuel consumption more than before – not just consumption on the road but also losses through stock shrinkage. Some of the 'other' reported interventions related to the improved monitoring of depot fuel stocks.

Conclusions

There is little doubt that the fuel blockades caused many problems for a freight and logistics industry that does not appear to have been very well prepared. Relatively few operators appeared to run out of fuel, but this would have been radically different had the crisis continued for just a few more days. The predictable reactions to the crisis were to change sources of fuel and to increase fuel stock levels, in some cases quite dramatically. The tactics used to circumvent the immediate crisis and the subsequent reactions both led to increased costs for hard-pressed operators and also resulted in reduced service levels for some of their customers.

Given the magnitude of the situation, it is perhaps surprising and a little worrying that many operators appear to have reverted to their pre-crisis practices regarding the purchasing and management of fuel supplies. A possible explanation, provided by some survey respondents, is the belief that both the industry and the government are now more alert to the situation, and will be ready to act to prevent a recurrence of the problems of Autumn 2000. Concern over the high costs of fuel stockholding is also apparent from the survey findings.

In general, the fuel price rises and the fuel crisis seem likely to have hit smaller operators the hardest. The survey strategy adopted was not amenable to the study of small operators, but it can be surmised that many will have run out of fuel during the crisis, due to greater dependence on refuelling 'on the road'. Many larger operators, as has been shown, managed to survive on their bulk supplies.

Larger well-established operators are also perhaps in a better position to afford the interventions that will yield the most significant fuel savings. The survey results reported in this chapter suggest that some operators are increasing their use of fuel saving interventions in response to the high cost of fuel. It is however clear that the scope for fuel saving is not unlimited. Some operators already make extensive use of such interventions, and there is a limit to how much more they can do. Against the background of the significant increases in the cost of fuel in recent times, it is clear that the trend towards fuel saving interventions is a notable case of an industry 'running just to stand still'.

References

ABC Freight Guide (published annually). London, Centaur.

Bamford, C.G. (1999). The Internationalization of Logistics in the UK. Chapter 21 in Waters, D (ed.), *Global Logistics and Distribution Planning: Strategies for Management, 3rd edition.* London, Kogan Page, 301-313.

Commercial Motor (2000). Top Hauliers are Making Less. *Commercial Motor*, 17-23 February 2000. Sutton (Surrey), Reed Business Information Ltd, p8.

Cooper, J. and Tweddle, G. (1994). Distribution Round the Clock. Chapter 13 in Cooper J (ed.), *Logistics and Distribution Planning: Strategies for Management.* London, Kogan Page, 195-206.

Coyle, M, (2000). "Reducing the Environmental Impact of Road Transport Operations: A Review of Interventions that can be applied by Fleet Operators". Paper presented at the CANTIQUE Workshop, Rome, 24-25 January.

Coyle, M. (2001). Reducing Costs through Driver Development. *Croner's Operational Costings Briefing*, Issue 18, 26 June. Kingston-upon-Thames, Croner Ltd, 2-4.

Coyle, M., Murray, W. and Whiteing, A.E. (1998). Optimising fuel efficiency in transport fleets. *Proc. Logistics Research Network Conference*, Cranfield University, 10-11 September, 255-264.

Department of Energy (1990). *Fuel Savings through Aerodynamic Styling of Trucks*. Energy Efficiency Demonstration Scheme Expanded Project Profile 335. London, Department of Energy.

Department of Energy (1992). *Fuel Savings Using Aerodynamic Styling On Articulated Trucks*. Energy Efficiency Demonstration Scheme Expanded Project Profile 355. London, Department of Energy.

Department of the Environment, Transport and the Regions (DETR), (1997). *Fuel Management for Transport Operators: Thorntons Plc*. Energy Efficiency Best Practice Programme Good Practice Case Study 342, London, DETR.

Fowkes, A.S., Firmin, P.E., Tweddle, G. and Whiteing, A.E. (2001). Freight Road User Valuations of Three Different Aspects of Delay. *Proc. European Transport Conference*, Stated Preference in Freight, Cambridge, 10-12 September, PTRC.

Motor Transport Operating Cost Tables. Sutton (Surrey), Reed Business Information Ltd (published quarterly).

Murray, W. (1999). Managing Agency and Temporary Drivers. *Croner's Road Transport Operations Bulletin*, 39, January, Kingston-upon-Thames, Croner Ltd, 1-16.

Road Haulage Association (RHA) (2000). *Fair Play on Fuel 2000*. Weybridge, Road Haulage Association.

PART III:
LEARNING FROM CRISES

11 Experiences and Lessons from Conducting an Emergency Survey of Motorists During the UK Fuel Crisis

GLENN LYONS AND MARK BEECROFT

Introduction

The transport research community is no stranger to conducting studies of travel behaviour. It has at its disposal a range of methodologies with which to gather behavioural data for analysis. Some approaches involve the study of ongoing 'normal' behaviour to gain insight into how and why people's daily routines, travel patterns and travel preferences and perceptions are formulated. Other approaches seek to create experimental environments to which individuals can be exposed in order to gather feedback on their reactions in such environments. Those environments might be artificial by exposing individuals to hypothetical situations. Stated preference techniques are a commonly used approach for this whereby individuals are asked to respond to a series of hypothetical choices (Ortúzar and Willumsen, 1998).

Alternatively, experimentation can intervene in real life. For example, May et al. (1998) studied drivers' responses to a range of road user charging (RUC) regimes. Participants from Newcastle University staff were provided with suitably equipped vehicles and a float of real money to use to go about their everyday lives with a hypothetical RUC scheme in place. This forced them to consider trade-offs between saving time on their daily journeys through making payments and saving the money they had been given. In Hampshire four companies took part in a teleworking trial in which employees were exposed for the first time to this way of working (Lyons, 1998). Travel diaries, questionnaires and interviews were used to acquire feedback from the participants. Such experimental arrangements typically afford the researchers substantial control over the experimental set-up. The experiment often has a clearly defined start and end point and it is common for two (or more) stages of feedback to be gathered from those taking part. The first stage assesses circumstance, attitudes and behaviour before the experiment is initiated. The second stage assesses comparable issues after the experiment is initiated. A subsequent stage may assess such issues again once the experimental environment has been removed. Such studies, by virtue of their detailed nature, can be expensive to set up and typically involve a comparatively small number of individuals.

The September 2000 fuel crisis in the UK inadvertently created conditions that were comparable to an experimental environment on a national scale involving huge numbers of participants. The 'experiment' involved the effective rationing of the supply of fuel to car users thereby rendering car use a finite resource. In these circumstances people were forced to confront their car dependence. They were required to review and appraise their daily routines and travel patterns and use of alternatives to the car.

It would have been impossible to intentionally create by design and then control the conditions encountered during the fuel crisis in the form of an experiment. There was no advance notice provided that the 'experiment' was taking place. Nevertheless the fuel crisis created an unprecedented opportunity to gather data that might yield fresh insight into car dependence.

This chapter presents and discusses the experiences of one group of researchers that undertook to monitor and evaluate the 'experiment'. The Transportation Research Group (TRG) at the University of Southampton conducted a survey of car users in the immediate wake of the fuel crisis to capture the experiences of individuals during the crisis in terms of its impacts on daily routines and travel. The process followed was not altogether untypical of a personal travel survey. However, the defining characteristic of this particular study was the time frame. Although the fuel crisis had no clearly defined beginning or end, the week commencing Monday 11 September 2001 in essence constituted the period of the fuel crisis. On Thursday 14 September the TRG made a decision to conduct a survey. By Friday 22 September (thanks to a major teamwork exercise) it had distributed 5000 mailback questionnaires. Colleagues elsewhere in England (in West Yorkshire (Leeds, Bradford, Wakefield, Kirklees, Calderdale), Leicester, Hertford and the London Borough of Hillingdon) distributed approximately a further 6000 questionnaires.

The chapter highlights the obstacles, opportunities and challenges faced in pursuing the survey. It also reflects on issues that contributed to its eventual success and those that served to compromise the survey's achievements. The validity and implications of referring to the fuel crisis as an experiment are considered further. The extent to which the research community might have been better prepared to seize this opportunity and how it might take preparatory steps now to capitalise on future opportunities is explored. The findings of the survey have been presented in Chapter 7 (Chatterjee and Lyons, 2002). Chapter 14 (Bonsall, 2002) will extend the discussion concerning what can be learned from crises.

At the end of this chapter the first author's diary of events is included. This provides a chronological account of the work undertaken in carrying out the survey. It also seeks to convey the pressures faced, the multitude of tasks to be completed and the importance of teamwork in the course of the survey process.

Survey Design

Survey Instrument

The principal aim of the survey was to understand how and why choices concerning trips and activities had been affected by the fuel crisis. To understand how trips and activities are affected by a change in circumstance a travel and/or activity diary is often an appropriate survey instrument to use. It allows an individual or household to document a chronological account of their routine over a number of days. However, with no advance notice there was no possibility of using travel diaries to record people's routines before the crisis to allow comparison with their routines during the crisis. In addition it was considered too late by the time of the week following the fuel crisis to expect individuals to retrospectively complete a diary for the previous week. A common concern with the use of diaries is that of underreporting of trips or activities by participants - diary surveys demand a lot of time of respondents (Kalfs and Saris, 1997). This consideration would have been an even greater concern in a retrospective context.

Key to selecting and designing an appropriate survey instrument for the fuel crisis was the need to act quickly to gather information from the public before their memories faded or became distorted. In acknowledgement of the retrospective nature of the survey, a lack of a complete and precise recollection of events and detail was considered inevitable. Therefore it was decided that rather than trying to acquire details of all individual trips made and activities undertaken by survey respondents, it was more appropriate to detect types of changes to routine and choices that took place and the incidence of such changes across respondents.

Three options for the survey instrument then existed, namely: (i) telephone interviews; (ii) face-to-face questionnaire interviews; and (iii) self-completion mailback questionnaires. Options (i) and (ii) faced three main limitations:

- *The inability to record detailed accounts from respondents*: These survey instruments are more amenable to closed questions with tick-box responses. In addition, oral surveys and telephone interviews in particular have been found to 'produce relatively poor results in terms of trip reporting in comparison with written surveys (e.g. mail-back travel surveys)' (Hassounah et al, 1993).
- *The need for trained personnel to conduct the interviewing*: The short timescales of the survey precluded either outsourcing the interviewing to a specialist agency or obtaining and then training a sufficient number of internal staff to perform the task.
- *The need to acquire a sufficient number of responses*: The unique nature and scale of the crisis suggested an inherent value in securing feedback from as many people as possible to harness the potential diversity of circumstance and behaviour faced during the crisis. These survey instruments would not have allowed large numbers of responses to be acquired within the time and resource available.

Option (iii) provided a greater flexibility in terms of question design and would allow a potentially greater number of individuals to respond with detailed accounts of their experiences and reactions. This option was therefore adopted.

Questionnaire Design

The final version of the resulting questionnaire is included at the end of the chapter. The main content of the questionnaire centred on four journey types: the journey to/from work; journeys made whilst at work; escorting children to/from school or pre-school; and grocery shopping. To avoid an excessive questionnaire length it would not have been feasible to explicitly consider more than four journey types. The types selected reflect a substantial proportion of travel undertaken in the UK. For 1997/99, the National Travel Survey identified that 'commuting', 'business', 'escort education' and 'shopping' journeys accounted for 54 per cent of all journeys made as car/van drivers and 33 per cent made as car/van passengers (DETR, 2000). A section of the questionnaire was devoted to each journey type with a common format for each section. (Other sections concerned personal details of the respondent, other changes to household routine and travel and the respondent's views about the fuel crisis.) Closed questions were asked to determine how each journey type is 'normally' undertaken. Respondents were then asked to describe in their own words any changes to their 'normal' practice that occurred during the fuel crisis week and how good or bad experience of the change was. Questions were also asked to ascertain whether such alternative approaches had been adopted previously or would be considered again beyond the crisis. Reference to 'normal' was for practical purposes in the questionnaire design. It is acknowledged that 'normal' for many people may not represent single, repetitive behaviour - they may have a lot of 'normal' variability in terms of the timing, location and mode of access for activities. In other circumstances a travel or activity diary would have been effective in capturing this variability.

Designing the questionnaire in this way raised some concerns. Firstly, the questionnaire would need to accommodate a range of different types of respondents facing different circumstances. For example, the patterns of activity and travel of retired people are likely to be substantially different to those of young, single, working people which will be different again to those of couples with young children. Using the same questionnaire for all sectors of the population of interest introduces elements of redundancy and irrelevance into the questionnaire design and restricts the opportunity for more targeted, relevant questions. In other circumstances it would have been appropriate to use an initial set of screening questions to stratify respondents. Each respondent could then have been provided subsequently with a variant of the general questionnaire designed more specifically for the corresponding stratum. Time constraints did not allow for this option.

A second concern related to the resulting length of the all-purpose questionnaire design. With its coverage of issues and requirement for free text responses to many questions it posed a risk of taking too long to complete. There was concern that this could have an adverse effect on the response rate for the survey. A calculated risk was taken. It was assumed that the length of time to complete the questionnaire would be offset by the high level of public interest in the fuel crisis and a wish amongst many to share their views and experiences by responding to the questionnaire. An additional financial incentive in the form of entry into a cash prize draw was also included. A reduced questionnaire length would have diminished the richness of the dataset emerging from the survey. In the event this judgement appears to have been reasonable - survey response rates are discussed later.

A major consequence of having many free-text response questions was the substantial time required subsequently to firstly transcribe the questionnaire responses and secondly process the responses and categorise them to allow quantitative analysis to proceed.

Piloting

For any questionnaire survey there is a need to ensure that the survey design, and in particular the questionnaire content, will work with those to whom the questionnaire is directed and will yield the data required. To do this requires pilot work whereby survey design is tested and refined - ordinarily this can be a lengthy process (Oppenheim, 1998). A questionnaire survey conducted without pilot work runs a serious risk of at least partial failure in terms of questions being misunderstood by respondents, key questions being omitted or response rate being very low because of the overall questionnaire design or means of distribution. When conducting an emergency survey no time is afforded for thorough pilot work. This places considerable pressure on those designing the survey. A limited attempt was made to pilot the draft questionnaire via some 10-12 friends, family and neighbours of TRG staff.

Questionnaire Distribution

It was determined at an early stage that the population of interest for the survey would be car drivers as opposed to the general public at large. A more precise definition of the population of interest that was targeted in practice was car users (thereby encompassing some individuals who do not drive but who, as passengers, depend to some extent upon the use of a car). The survey was not designed with the intention of learning how those who do not use a car were affected by the crisis. This is not to say that they were not affected – overcrowding on public transport services as patronage levels swelled may well have caused disruption or inconvenience to non car users. Indeed Thorpe et al. (2002) found that, during the crisis, some individuals switched from public transport to car use perhaps because of a deterioration in the level of service on public transport and more free flowing roads due to reduced traffic levels.

One option for distributing mailback questionnaires was to use the postal service to deliver questionnaires to households in specified post code areas. However, this had two distinct disadvantages. Postal distribution introduces a wastage given the inability to target only households with individuals within the population of interest (i.e. car users). Using the postal service for distribution would have taken a number of days to arrange, causing unacceptable delay. Further to this, the postal service itself was suffering at the hands of the crisis. Five alternative forms of distribution were actually utilised:

- *Fuel stations.* With the crisis itself over and tankers replenishing fuel stocks on forecourts, fuel stations seemed to provide a natural distribution point. The majority of individuals visiting fuel stations are, by implication, car drivers or users (notwithstanding the increasing attempt by many stations to also act as local supermarkets). At the point of visiting a fuel station, refuelling and the issue of the crisis were likely to have been uppermost in people's minds. A number of fuel stations agreed to have questionnaires placed on their forecourts.
- *Primary schools.* Primary school children by implication are in households with young families and most, if not all, will involve parents in escorting the children to school. A number of schools agreed to send a questionnaire home with each child for their parents. Whilst this would not necessarily have only targeted the population of interest it ensured that the survey data would have school escort trip information well represented, subject to a good response rate. This form of distribution, as with that above, also minimised the manpower requirements of distribution imposed on the research team.
- *City and town centres.* City and town centres provide a means to target a broader cross section of the public, an opportunity to target distribution at the population of interest and a greater likelihood of targeting individuals who will complete the questionnaire given the voluntary nature of receiving a questionnaire. Within this form of distribution, car parks and shopping areas were targeted. This form of distribution required a number of staff to support it. Depending on the exact point of distribution this can also be more susceptible to weather conditions (which were not altogether favourable during the survey) than other methods.
- *Door-to-door.* Door-to-door distribution is equivalent to the postal service option discussed earlier, i.e. questionnaires are delivered to households without ascertaining either whether the household contains individuals within the population of interest or whether such individuals are prepared to complete a questionnaire. Due to unfavourable weather conditions that limited the rate of distribution on-street, door-to-door distribution was used as a supplementary measure and targeted at villages.
- *Electronic distribution.* In addition to the distribution of the mailback questionnaire, a duplicate version of the questionnaire was created for the web. This allowed individuals with access to the Internet from any physical location in the UK to participate in the survey. A key

determinant of success for this form of distribution is the degree of awareness of its existence amongst the population of interest. This is discussed later in this chapter.

The TRG is based in Southampton. For mailback questionnaire distribution in the forms described above, the geographical coverage of the survey was limited to Hampshire and South Wiltshire. There was concern that such a limitation would substantially impinge upon the extent to which the survey response sample would be representative of the UK, or England, as a whole. The Internet proved an invaluable aid to addressing this problem, separate to its use for the web-based questionnaire. The chapter authors share in the operation of an e-mail discussion list for young transport professionals. A request was made to list members from across the UK for support in questionnaire distribution. A number of colleagues agreed to help in various parts of England as indicated in Figure 11.1 below. The questionnaire was produced using Microsoft Word. Therefore it was possible to email the document to colleagues. They in turn produced sufficient paper copies for their own distribution needs (in the case of Leicester the Word questionnaire document was distributed electronically with respondents either completing and returning it electronically or printing it out for completion and returning it by post).

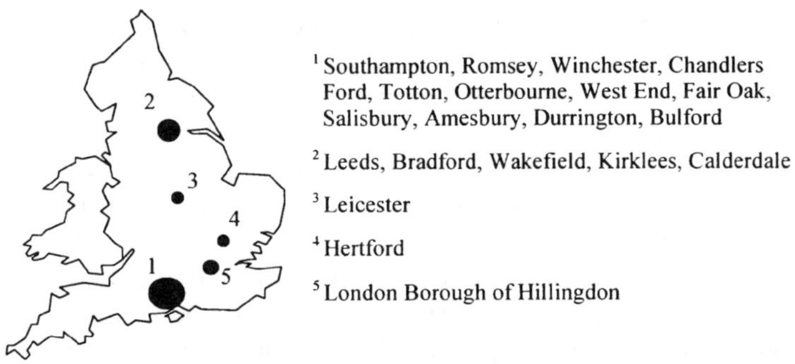

Figure 11.1 Areas of distribution for the mailback questionnaire

Without the Internet it would not have been possible to expand the geographical coverage in this way. Nevertheless, under the tight time constraints for the survey, the research team at Southampton was unable to record survey distribution information for areas 2-5 to the same level of detail as for area 1 (available details concerning areas 2-5 are included at the end of this chapter[1]). For areas 2-5 an area identifier only was stamped on each questionnaire distributed. For area 1, each questionnaire was stamped with a unique ID number. The specific location and method of distribution was logged for nearly all ID numbers.

Response Rates

Table 11.1 shows the estimated response levels and rates attained according to the distribution methods employed for area 1. Table 11.2 shows the response rates for the five survey areas.

In terms of the four principal means of distributing the mailback questionnaires the highest response rate was from petrol stations (average response rate approximately 25 per cent), followed by town/city centres (22 per cent), house to house (20 per cent) and primary schools (19 per cent). The comparative response rates are intuitively sensible. For petrol stations and town/city centres, all individuals have 'volunteered' to take a questionnaire whereas for primary schools and house to house individuals receive a questionnaire regardless. What is perhaps more surprising however is that the response rates for 'volunteers' are not substantially greater than for others and yet town/city centre distribution was much more resource intensive. This might provide a lesson for any future 'crisis' survey where time and resources are at a premium. Nevertheless, response rates are clearly not in themselves a measure of the success of the survey.

The response rate for area 1 is noticeably higher than for the other areas as shown in Table 11.2. There are two probable reasons for this. Firstly, the figures for how many questionnaires were distributed in areas other than area 1 are approximate and represent maximum values. Many of these questionnaires may have been despatched for distribution without actually being taken by prospective respondents. Secondly, as described below, the survey was well publicised in area 1 whereas in other areas publicity was limited or nil.

Table 11.1 Comparison of response rates across distribution methods for area 1

Location	Distribution means	Distribution	Responses	Response rate (%)
Durrington[3]	Primary school (infant and/or junior)	145	36	24.8
Chandlers Ford[9]		200[a]	41	20.5
Winchester[5]		204[b]	-	-
West End[10]		132	27	20.5
Romsey[6]		226	28	12.4
Bulford[3]	House to house	170	30	17.6
Durrington[3]		144	34	23.6
Otterbourne[11]		20	3	15.0
Chandlers Ford[9]		200	46	23.0
Fair Oak[8]		100	13	13.0
Amesbury[2]	Town or city centre	105	25	23.8
Salisbury[1]		105	22	21.0
Southampton[4]		600	89	14.8
Romsey[6]		726	181	24.9
Winchester[5]		600[c]	143	23.8
Durrington[3]	Petrol station	110	34	30.1
Romsey[6]		288	80	27.8
Totton[7]		300[a,d]	35	11.7
Winchester[5]		208	51	24.5
Durrington[3]	Rangers petrol station - account holders	100	19	19.0
Southampton[4]	University staff club	61	19	31.1
Winchester[5]	Alternative transport day stall	136	37	27.2
Winchester[5]	Email/web[e]	unknown	21	unknown

Table 11.1 Comparison of response rates across distribution methods for area 1

Location	Distribution means	Distribution	Responses	Response rate (%)
Amesbury[2]	Health Centre / Doctor's surgery	20[a]	4	20.0

1 Salisbury is a Cathedral City in Wiltshire with a population of approximately 40,000

2 Amesbury is a small town in Wiltshire 8 miles north of Salisbury with a population of approximately 6,000

3 Durrington and Bulford are villages in Wiltshire 1 and 2 miles from Amesbury respectively

4 Southampton is a City in Hampshire with a population of approximately 220,000

5 Winchester is a Cathedral City in Hampshire with a population of approximately 35,000

6 Romsey is a town in Hampshire 8 miles north-west of Southampton with a population of approximately 13,000

7 Totton is a town in Hampshire 4 miles west of Southampton with a population of approximately 28,000

8 Fair Oak is a village in Hampshire 8 miles north-east of Soutampton

9 Chandlers Ford is a village in Hampshire 6 miles north of Southampton and adjacent to the town of Eastleigh

10 West End is a village in Hampshire three miles east of Southampton

11 Otterbourne is a village in Hampshire 9 miles north of Southampton and adjacent to the town of Eastleigh

a Not all questionnaires made available for the public may have been taken/distributed - the response rate is therefore a minimum estimate

b No responses at all were received from this batch - it is assumed therefore that they were not distributed

c Approximately 50 of this batch were distributed house to house

d No responses from the latter 100 assumed to be distributed were received - it assumed that these were not therefore distributed

e The questionnaire as an electronic Word document was emailed around an organisation in Winchester

Table 11.2 Comparison of response rates by survey distribution area

Distribution area	Distributed	Returned	Response rate (%)
1 - Hampshire/Wiltshire	4596	1018	22
2 - West Yorkshire	5000	416	8
3 - Leicester	n/a	77	n/a
4 - Hertford	760	70	9
5 - Hillingdon	500	52	10
unknown	n/a	26	n/a
All areas*	10856	1659	15

* Distribution values for areas 2, 4 and 5 are approximate and n/a cells are treated as zero values - hence values for all areas are approximate

The Web as a Questionnaire Medium

The UK has a population of nearly 60 million (Office for National Statistics, 2001). Estimates of the number of people in the UK with Internet access vary and the actual figure is continually growing. At around the time of the fuel crisis one source put the number of people in the UK using the Internet every month from home at 11 million in September 2000 (Jupiter MMXI, 2000). In addition to home users there are others who have access at home but do not make use of it and others who use the Internet at work. In other words a substantial proportion of the UK population would have been able to access a questionnaire placed on the web. This represented an opportunity to potentially harness questionnaire responses from literally thousands of people across the UK at very little cost.

A web questionnaire is essentially an on-line form. An individual accesses the questionnaire using a web browser and completes it electronically using the computer keyboard and mouse. Finally an on-screen button is clicked to submit the response electronically back to the computer hosting the web questionnaire. Responses are automatically stored in a database removing the requirement for transcription.

However, one factor is decisive in the effectiveness of a web questionnaire – individuals must be aware of the existence of the questionnaire and easily be able to locate it if they are to become survey respondents. Therefore, for the web version of the fuel crisis survey (identical to the mailback questionnaire) it was essential to promote its existence. Promotion was pursued using a variety of means:

- messages sent to email discussion lists (with encouragement for such messages to be forwarded to others in a chain letter style);
- requests to and offers from other web site operators to place links to the questionnaire on their web sites; and
- radio and press coverage.

At a local level (the Southampton area) all these means of promotion were achieved. Regrettably, at a national level attempts to gain radio and press coverage failed. A prime time evening slot was secured for a promotional interview on BBC Radio 2 on Wednesday 20th September. However, minutes before the broadcast the show's producer withdrew the opportunity because of fears triggered by another radio station (Cardiff based Red Dragon FM) that the fuel crisis was recommencing (BBC, 2000). It will never be known how many responses could have been generated from this crucial promotional opportunity. Nevertheless, over 700 responses were received for the web questionnaire during the 12-day period for which it was made available. Figure 11.2 shows the number of daily responses received. It also highlights the points at which different forms of publicity became effective.

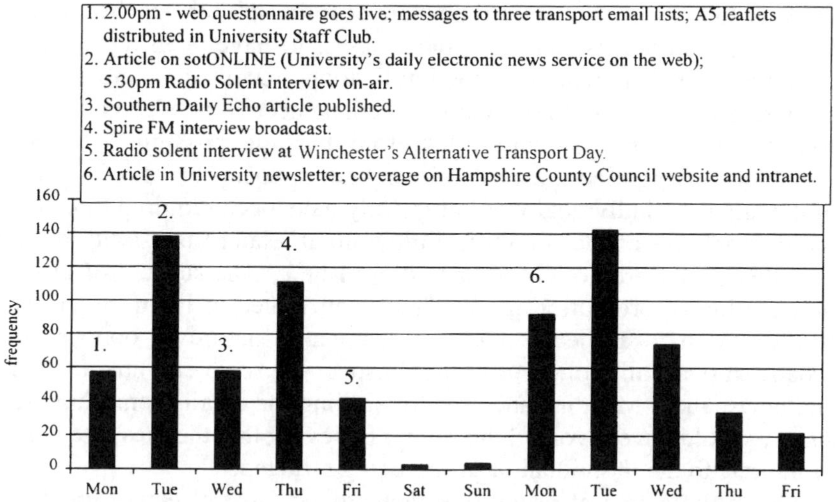

Figure 11.2 Daily responses to the web questionnaire between Monday 18th September and Friday 29th September 2000 and corresponding known instances of promotion

In addition to the questionnaire completion, web survey respondents were asked to indicate how they had learned of the web survey. From responses to this open question it is clear that the most effective means of promotion was email. Approximately half of the respondents had learned of the web survey from receiving an email message. From indications of the means of promotion attracting respondents, the chain letter style of promotion did indeed take place. The original promotional email messages posted to discussion lists by the TRG were forwarded to a number of other discussion lists, circulated within organisations and passed around between friends and colleagues. On the second Monday and Tuesday (25th and 26th of September) 234 responses were received. Of these, 53 per cent arose as a direct result of promotion item 6 shown in Figure 11.2. Promotion item 6 represents promotion within the workplace and in this case involved organisations with widespread Internet access amongst their staff. Local radio and newspaper articles only achieved a minor effect in terms of successful promotion.

These observations highlight the effectiveness of the 'electronic grapevine' in promoting a web survey. There are two suggestions regarding the cause of this. Firstly, there was almost universal interest in the topic of the survey across the UK and a consequent willingness and enthusiasm for people to pass on promotion of the survey. This would be unlikely to exist in ordinary circumstances for a web survey. Secondly, this form of promotion is likely, in most cases, to have been received in the workplace environment and, by implication, received in most cases by individuals sat at a computer with immediate access the web and hence the web survey. The promotion therefore has the possibility to achieve immediate effect in contrast to radio and newspaper promotion. In the case of the latter the individual would typically have been required to note down the web address of the survey to follow up at a later time - with the initial stimulus of the promotion perhaps diminished. The success of electronic propagation of promotion might also be attributed to its more narrowcast (transmission to a specified list of recipients) nature as opposed to the broadcast (transmission to multiple, unspecified recipients) nature of radio and newspaper. Most instances of forwarding the email promoting the web survey would have involved the sender believing that the message would be of interest to the individual or group being targeted.

Given the forms of promotion that proved most effective for this web survey, respondents are likely to have been mostly professional, in contrast to the mailback survey which captured a broader cross-section of the population. Hence response bias issues must be addressed. Chapter 7 focuses on the mailback survey responses. The analysis of the web survey is not included in this book. It will be reported elsewhere in due course along with a comparison of the two sets of responses.

There is a growing experience within research communities of using the Internet for conducting surveys (e.g. Polak et al., 1999). As populations become increasingly familiar with and able to use the Internet then it is likely to become an increasingly valuable survey instrument. This is particularly the case where large numbers of people (possibly in disparate locations) are affected by a common event or crisis which may be short lived. A cluster of unannounced crises hit the UK in the winter of 2000/2001. Following the fuel crisis, a major rail crash resulted in a huge infrastructure programme causing serious disruption to passenger train services. Flooding affected large parts of the UK and foot and mouth returned to the UK for the first time since 1967. Each of these in different ways (whilst none of them were as short lived as the fuel crisis) provided opportunities to learn more about human behaviour in a climate of intense media interest capturing national attention – fruitful territory for the use of an Internet survey.

Funding Research in a Crisis

A substantial proportion of academic research in the UK is funded by research councils. A typical process to gain funding for a research proposal from the Engineering and Physical Sciences Research Council (EPSRC) is as follows. A research proposal is submitted to the EPSRC and sent for review to individuals with recognised expertise in the field. This can take several weeks. If reviewers' feedback is favourable then either: (i) funding may be awarded if the value of the proposal is small; or (ii) the proposal joins a number of others that are taken forward to a Panel meeting which may take place some further weeks hence. In the case of the latter, proposals are ranked and a funding 'cut-off' identified which determines which projects will be funded and taken forward. To varying degrees all research sponsors follow similar procedures of assessment that usually preclude any immediate decision on a request for funding.

For the fuel crisis it was considered vital that the survey commenced without any delay. The TRG therefore took the unusual step of choosing to underwrite the cost of conducting the survey in case funding from elsewhere could not be secured subsequently. This enabled the survey work to be commenced without delay. Three working days into the survey a short proposal for funding was prepared and submitted to the EPSRC. The EPSRC demonstrated commendable flexibility and haste in its response. Upon receipt of the proposal to cover the costs of the survey, EPSRC was able to agree to fund the work within the space of just over two weeks having sought and received comments from three referees and avoided passing the case to a Panel meeting. The TRG then, having completed the survey itself, submitted a follow-on proposal to the EPSRC for funding to conduct the data entry, analysis and reporting for the survey. This proposal was refereed in the normal way and awarded funding.

This approach to securing funds for the work proved highly effective. It enabled the time critical nature of the survey work itself to be addressed to capture the research data. It was then possible in a more measured way to take stock of the outcome of the survey and the number of responses and to consider a suitable approach for establishing and disseminating research findings from the collected data. However, the success of the survey relied upon the research group itself undertaking to commence the work with no guarantee of funding. To wait even the two weeks for approval from EPSRC would have seriously compromised the outcome of the work undertaken.

Teamwork

The success of the survey was crucially dependent upon the large number of individuals within and associated with the TRG and those in other parts of the UK who contributed to the work. As a reminder, in the space of five working days the TRG team, from scratch: initiated the project; designed, tested and revised a mailback and web questionnaire; distributed 5000 copies of the mailback questionnaire; and achieved publicity for the web questionnaire drawing over 360 responses in its first four days of existence. A multitude of tasks was involved alongside the survey design including: the mundane process of preparing copies of ID stamped questionnaires with mailback envelopes and preparing laminated posters for use by the questionnaire distributors; technical matters associated with the web questionnaire; publicity efforts; recruitment of temporary staff; and distribution coordination.

The TRG is comprised of some 35 staff and postgraduate research students. The majority of these took a role, great or small, in the survey. In addition, friends and family were drafted in to provide temporary support. This short-lived and intense project arose with no prior notice. No staff had the opportunity to forward plan in order to accommodate their contribution to the fuel crisis survey alongside their other work commitments. Business as usual in the group still had to continue. It is therefore a credit to all those concerned how effectively a large team of people worked together to achieve the goals of the project.

An Experimental Environment?

Earlier in this chapter the fuel crisis was referred to as a national 'experiment'. In many regards this description is fitting although it is prone to be misleading. Referring to the fuel crisis in this way wrongly implies a degree of experimental control. In interpreting the results of any experiment and judging what the implications might be, it is important to ensure that there is no ambiguity with regard to how the experimental results have arisen and what they represent.

It would be wrong to suppose that for all the fuel crisis survey respondents, the same degree of fuel rationing applied or that the fuel crisis began and ended at the same time for all respondents. In practice different parts of the UK suffered varying levels of fuel shortages at different times. Across the population some individuals secured ample fuel for their immediate requirements whilst others had little or none. The population as a whole could not be aware of the length of the crisis until it had concluded.

Therefore conduct across individuals is likely to have varied depending upon perceptions of likely duration as well as the changing degree of fuel availability. This highlights the added complications faced, over and above a more conventional experiment or survey, in conducting and interpreting the analysis of the data obtained.

It is impossible to resolve precisely how such considerations impinge upon the validity of the survey findings or their interpretations. In many ways this is true of any experiment. Crises hold the prospect of rich rewards in terms of the effects they can have on behaviour if monitored but such rewards come at the price of the caveats that must surround the results. Such matters for this particular survey are considered in greater detail in Chapter 7 and more broadly in Chapter 14.

Preparing to Monitor and Evaluate Behaviour in a Future Crisis

The fuel crisis survey met with a surprising degree of success given the constraints and pressures that it faced. In some respects it merely demanded a well-oiled execution of all the stages and principles of good practice in conducting any travel behaviour survey. In other respects it has exposed some of the unique challenges and opportunities that a national crisis can bring for a survey that seeks to assess its impacts. It might be argued that there is little to be gained in devising a contingency plan for monitoring and evaluating behaviour in a future crisis. Each crisis will possess its own unique characteristics. In spite of the spate of crises that has recently beset the UK, crises remain an exception rather than a norm. However, crises can represent tremendous research opportunities when they occur. They can create conditions akin to experimental environments on a large scale that behavioural researchers could only dream of creating themselves. It therefore seems appropriate that this chapter should conclude with contemplation of a contingency plan, or at least some points of discussion regarding one.

Funding is a dilemma if external sponsors are to be approached. A decision for the study of a crisis to go ahead may be required in a matter of days or even hours rather than the weeks or months typically required. A sponsor might have in place an allocation of emergency funds to support a crisis study. However, to release such funds it would not be sufficient to be persuaded of the merits of an opportunity presented. The research team would still need to be judged as credible to carry out the work. For these reasons it is suggested that the model that was followed for the fuel crisis survey is an appropriate one. It should be the internal responsibility of a research team to be prepared to initially meet the costs of a crisis study. If the team is confident of its value then it should be confident of recovering its costs from a sponsoring body. Sponsoring bodies should be encouraged to have in place a mechanism to deal with such a situation if they do not already do so.

This chapter has highlighted the pressures that survey design can be placed under. Time is of the essence and a multitude of decisions and tasks must be addressed. Under pressure it can be easy for such matters to be addressed in a sequential fashion rather than in parallel. Thinking in parallel affords at least a few hours or days of advance planning for matters such as pilot work. In the case of this survey, only limited pilot work was possible because of no advanced planning. Ultimately a crisis survey faces the limit of the number of available person hours that can be devoted to the work. However, the availability, at the work commencement, of a checklist for all decisions and tasks can enable a project plan to be drawn up and the critical path of activities to be identified. This is normal project management procedure, but can be easily overlooked or inadequately addressed in crisis conditions.

Survey design should consider the extent to which the crisis in question can be defined in terms of experimental design. Thinking in this way can expose some of the uncontrolled factors that are at play. Whilst control cannot be imposed, questions can be posed that will minimise the risk of ambiguity or inability to interpret the findings from the experiment.

Teamwork was instrumental in the success of the fuel crisis survey. To carry out the work successful not only required sufficient people but an ability to match tasks and decisions to individuals' abilities and expertise. For a future crisis survey this would principally include those within the organisation pursuing the crisis survey. However, prompt identification of other individuals and contact details for matters such as issue of press releases, obtaining police permission for on-street survey distribution, recruitment of temporary staff and mass production of questionnaire material is also important. It may be advantageous for more than one research organisation to join forces to address a crisis survey. This will require an even more thorough appreciation of the roles of different team members. It is suggested that either within a research organisation or within a research community a crisis task force be identified which can be mobilised when required. This might consist of a number of individuals and their corresponding skills identified for inclusion in a potential crisis survey research team. More likely is that it would consist of one or more identified individuals who possess the knowledge and organisation skills to advise on or take a direct role in the management of a crisis survey.

There is still much to learn about the use of the Internet to conduct surveys (particularly in terms of the self-selection of respondents and the associated bias in responses) but it is a medium that is increasingly familiar to growing proportions of national populations. As discussed earlier in this chapter its principal advantage is being able to make a questionnaire widely available in a very short space of time when only limited survey resources are present. For the fuel crisis survey the Internet questionnaire was considered a supplementary means of gathering information to the main mailback survey. It is suggested that for a future crisis a reversal of this priority might be considered (particularly at such a time when Internet access becomes sufficiently widespread to ensure a representation of the entire population in survey responses). The public increasingly look to the Internet to find and exchange information. A survey should be conducted in a way that takes advantage of this. For example, during the Foot and Mouth crisis it is estimated that nearly 500,000 Britons visited the Ministry of Agriculture, Food and Fisheries website in March 2001 seeking information (Jupiter MMXI, 2001). Imagine the potential promotional opportunity a link on this website would have presented for a crisis survey.

A crisis survey is not something to be undertaken lightly. It is extremely demanding. One moves quickly from observing or being part of the crisis itself to being part of a newly created crisis - namely the research crisis. The plus side is the prospect of securing a valued contribution to knowledge that would otherwise have been lost forever.

Acknowledgements

Sincere thanks are due to all those individuals who kindly helped in this venture. They include: members of the TRG and their partners and friends; Reg Harman (Independent Consultant and Visiting Senior Research Fellow with the TRG); Andy Salkeld (Leicester City Council); Jeanette Sargent (West Yorkshire Passenger Transport Executive); Mark Silverman (London Borough of Hillingdon); and Paul Chu (Mott MacDonald). There are numerous others, many of whom we are not even aware, not least those who shared in the propagation of the web survey promotion and those who considered our request for funding. To everyone concerned we owe a debt of thanks.

Notes

1 Known details concerning forms of distribution in areas 2, 4 and 5 are as follows. *Area 2*: approximately 3,000 distributed through local councils, to local councils' staff and community Groups in Leeds, Bradford, Calderdale, Kirklees and Wakefield (proportions of 'public' and 'internal' distribution are not known); and an additional 2,000

questionnaires were distributed in city centre car parks in Leeds. *Area 4*: 600 to Simon Balle Comprehensive School in Hertford, 100 in Gravesons Department Store in Hertford and the Tourist Information Centre, 60 directly by hand. *Area 5*: 400-500 distributed at petrol stations, libraries, housing department and voluntary user groups (approximately 100 each).

References

BBC News Online. Website. Careless talk costs litres. Available at: http://newsvote.bbc.co.uk/low/english/uk/wales/newsid_933000/933405.stm [accessed 04/09/01].

Bonsall, P. (2002). What can be Learned by Studying Transport Crises? In Lyons, G. and Chatterjee, K. (eds.) *Transport Lessons from the Fuel Tax Protests of 2000*, Ashgate, Aldershot.

Chatterjee, K. and Lyons, G. (2002). Travel Behaviour of Car Users During the UK Fuel Crisis and Insights into Car Dependence. In Lyons, G. and Chatterjee, K. (eds.) *Transport Lessons from the Fuel Tax Protests of 2000*, Ashgate, Aldershot.

DETR (2000). *Transport Statistics Great Britain: 2000 Edition*. Department of the Environment, Transport and the Regions, London, TSO.

Hassounah, M.I., Cheah, L.-S. and Steuart, G.N. (1997). Underreporting of trips in telephone interview travel surveys. *Transportation Research Record 1412*, TRB, 90-94.

Jupiter MMXI (2000). 11 million people in the UK now use the Internet every month. Press release, November. Available at: http://uk.jupitermmxi.com/press/releases/20001114.jsp. [accessed 03/09/01].

Jupiter MMXI (2001). Britons turn to the Internet in a crisis. Press release, April. Available at: http://uk.jupitermmxi.com/press/releases/20010417.jsp. [accessed 03/09/01].

Kalfs, N. and Saris, W.E. (1997). New data collection methods in travel surveys. In: Ettema, D.F. and Timmermans, H.J.P. (eds) *Activity-Based Approaches to Travel Analysis*, Pergamon.

Lyons, G.D. (1998). An assessment of teleworking as a practice for travel demand management. *Transport*, 129(4), ICE, 195-200.

May, A.D., Bonsall, P.W. and Hills, P.J. (1998). Evaluation of driver response to road user charging systems. *Proc. 9th International Conference on Road Transport Information and Control*, April, London, IEE, 30-34.

Office for National Statistics (2001). *Mid-2000 UK population estimates*, August, National Statistics, London.

Oppenheim, A.N. (1998). *Questionnaire Design, Interviewing and Attitude Measurement*. Pinter.

Ortúzar, J. de D. and Willumsen, L.G. (1998). *Modelling Transport*. Wiley.

Polak, J., Plaxton, J., Jackson, P.G. and Wofinden, D. (1999). Using the Internet to improve long-distance travel data collection. *Transportation Research Record 1660*, TRB, 148-155.

Thorpe N., Bell, M., Polak J. and Noland R. (2002). A Telephone Survey of Stated Travel Responses to Fuel Shortages. In Lyons, G. and Chatterjee, K. (eds.) *Transport Lessons from the Fuel Tax Protests of 2000*, Ashgate, Aldershot.

DAILY DIARY OF G. LYONS

DAY 1 – Thursday 14th September 2000

At home teleworking to conserve fuel. My wife has walked the children to school.

On Tuesday I suggested to an email discussion list that the fuel crisis was an opportunity to understand how people react when they have to compromise their car dependence. This would be a research opportunity if not for the short notice and the problem of funding.

Had a telephone conversation with Professor McDonald, our Group Director, to discuss any possibilities for pursuing such research. We agreed the Transportation Research Group itself would underwrite the costs of an 'emergency' survey in the absence of an immediate source of funds.

Late today there were signs from media reports that the immediate problems of the fuel crisis the blockade of refineries and distribution centres might be about to end. A survey to capture the effects of the crisis will need to be conducted as soon as possible. A mailback questionnaire will be the best option this will need to be ready for distribution early next week. I called Mark Beecroft at the office and we agreed to assume joint responsibility for completing the survey. Mark has suggested a TRG brainstorm meeting first thing tomorrow.

I started thinking through concepts and ideas for the questionnaire and discussed them with my wife as a prospective respondent. I also drew up an action list for tomorrow. Mark, meanwhile, plans to ring round family and friends to get insights into how the crisis has been affecting them.

DAY 2 – Friday 15th September 2000

At work just after 6am to continue with questionnaire design focussing on a section for each main trip type which can identify typical travel arrangements and changes (if any) during the crisis. Copies of partially complete draft questionnaire given to those attending the brainstorm meeting.
Six staff attended the meeting. The main point of discussion was the detail of the survey instrument. We considered travel diaries, interview surveys and preliminary screening surveys but time constraints necessitated the more generalised mailback questionnaire approach. Open versus closed questions and the need to capture interdependencies between different trips and activities were discussed. It was agreed to target the survey at adult car users only. 3000 questionnaires was estimated as a target for distribution. I undertook to complete a draft of the questionnaire by the end of the day. Meanwhile Mark, with support from three other staff, set about preparing 3000 A5 envelopes with FREEPOST address labels.
With concern over geographical coverage, Mark and I contacted a number of colleagues in other parts of the UK who agreed to help by undertaking their own distribution of the questionnaires. Notably, Jeanette Sargent of the West Yorkshire Passenger Transport Executive undertook to make some enquiries with the relevant local authorities in and around Leeds.
EPSRC contacted me about other matters. We discussed what the TRG was doing and they suggested if I could get a proposal for funding to them immediately they could turn it around very promptly subject to a favourable responses from referees (a similar exercise had been completed in response to an earthquake when researchers had needed funds to travel to the zone to collect data/information). I explained that there was simply not time at this stage to prepare a proposal and complete the survey design.
I spent the rest of the day designing the questionnaire. Once it was completed it was emailed to all those in different parts of England who were interested in taking part in the distribution. As a substitute for piloting the survey, five members of staff agreed to take home copies of the draft questionnaire to test on a limited number of people.
At home this evening I spent several hours preparing an HTML version of the questionnaire to be posted on the TRG website.

DAYS 3 and 4 – Saturday 16th and Sunday 17th September 2000

I spoke with a local petrol station proprietor who agreed to distribute the questionnaire. The station has a typical footfall of 1000 people a day. It was anticipated that on Monday and Tuesday the garage would be very busy as fuel supplies arrived. With an estimate of 1 in 8 people taking a questionnaire over three days up to 500 questionnaires would be required for this point of distribution alone. Working on the basis of targeting perhaps three petrol stations and three schools as well as city centre distribution, it became clear that 3000 was an inadequate estimate of total distribution. I decided to revise the total to 5000 and prepared a detailed action list for Monday to ensure the survey could continue to plan.

Late Saturday evening I sent an email to the Radio 2 'Drive Time' show in the hope they would be interested in covering the survey.
My wife and her friend began making plans to distribute the questionnaire on Tuesday and Wednesday in Salisbury and Amesbury.
My wife , one of my neighbours and a family friend each completed the questionnaire and provided feedback. Work colleagues also emailed through feedback. I spent more of the weekend making notes on how to refine the questionnaire.

DAY 5 – Monday 18th September 2000

Arrived at work at 6.00am to begin revisions to the questionnaire. Following further emails received with feedback, the final version of the questionnaire was ready. A colleague did the proof reading and it was then dispatched at 9.30am for mass production.

Three research students were enlisted to help other TRG staff with preparation of more mailback envelopes and in turn putting the photocopied questionnaires together with envelopes. Each questionnaire was also stamped with a unique I.D. number. A colleague's wife joined the team and assumed responsibility for contacting schools and petrol stations to secure support for distribution.

The on-line web questionnaire was revised and two other TRG staff dealt with the technical tasks associated with setting up the questionnaire to allow on-line responses to be stored directly into a database without the need for transcription. After external testing the web questionnaire went live at 2.00pm. I then posted email messages to three email announcement/discussion lists promoting the web questionnaire.

I wrote a press release that then required liaison with the University's Public Affairs department to establish a final version which they would release. Attempts to get a supporting quote from a leading transport organisation failed, with no-one available to comment. Later in the day the press release was issued and was also placed on the Latest News section of the University website. A TRG colleague tried unsuccessfully to generate interest from four major newspapers in covering the survey.

I then prepared A4 colour posters (which required lamination), name badges and letters for the staff, family and friends enlisted to distribute the questionnaires on-street. With permission obtained, questionnaires and A5 promotional flyers for the web questionnaire were taken over to the University Staff Club. A cover letter to be attached to all the questionnaires being distributed at schools was written and mass produced. Meanwhile Mark contacted all the relevant police authorities for permission to distribute questionnaires.

I departed for home with sufficient questionnaires for the local infants school, petrol station and distribution in Salisbury the following day. Our TRG secretary also took sufficient questionnaires for distribution at her daughter's primary school.

Late that evening I checked my email only to discover two messages concerning failed attempts to submit the web questionnaire. I investigated and discovered the problem but was limited from home in being able to fully resolve it. I also found an email from BBC Radio 2 who had tried (too early) to access the questionnaire on the web I sent an email indicating that it was now available. Finally I went onto the EPSRC website and obtained the relevant documents for submitting a research proposal.

DAY 6 – Tuesday 19th September 2000

Arrived at work at 6.30am and wrote a proposal to EPSRC for the costs to cover the survey which was then agreed with the Group Director and passed to our secretary for submission to EPSRC.
Colleagues fully resolved the problem with the web questionnaire.
Mark spent the day co-ordinating and taking part in the distribution of the questionnaires in Romsey, Winchester and Southampton. Distribution staff got caught up in the new wave of panic buying of fuel triggered by the Welsh radio station Red Dragon.
The University's Public Affairs Department called to say they were receiving enquiries concerning the press release. The Southern Daily Echo was prepared to drop another article in the following day's issue to accommodate a 600 word piece. I wrote an article and passed it to Public Affairs. Public Affairs sent over a photographer for which I had to ask a colleague go and buy some razors to allow me to have an emergency shave! Then did a recorded interview with South City FM. Further phone calls with Radio 2 secured a live prime time 5 minute slot at 6.15pm that evening. Radio Solent called and a live interview was arranged on Drive Time at 5.30pm, also that evening.
The interview with Radio Solent went well including the clear mention of the Website address. Radio 2 then telephoned to say that the show's host was uncomfortable about doing the interview in view of the resumption of (unnecessary) panic buying this opportunity was lost.
Reports back from those distributing questionnaires indicated that rainy weather had not been helpful although interest in the survey was generally strong in light of the temporary resumption of panic buying some drivers even wound their windows down and beckoned to distributors to give them copies of the questionnaire to complete.

DAY 7 – Wednesday 20th September 2000

The Southern Daily Echo article was published today.
I was teleworking once again - no longer because of the fuel crisis but because of a work-related crisis caused by pursuing the survey I am responsible for a major industry workshop in London tomorrow concerning traveller information and had to prepare for the day including a substantial presentation to open the event.
During the day contact was made with Salisbury's Spire FM and I conducted a telephone interview.
Mark continued with the substantial task of questionnaires distribution.

ADDENDUM: Thursday 21st – Monday 25th September 2000

On Thursday an abridged version of the Spire FM interview was broadcast hourly during the day but unfortunately without the web address included.
By the end of Thursday, the TRG team had distributed 5000 questionnaires.
Alongside the frenetic activity locally, colleagues in Leeds, Bradford, Wakefield, Kirklees, Calderdale, Leicester, Hertford and the London Borough of Hillingdon has been similarly engaged in the distribution of a further 5000-6000 questionnaires.
Alongside the organisation of distribution, Mark attempted to log the details of activities and to arrange payment of casual staff.
The Southern Daily Echo article yielded an invitation, which Mark accepted, to have a stand at the Winchester Alternative Transport Day, part of the European Car-Free Day campaign, on Friday. This also resulted in a radio interview with BBC Radio Solent within which the web address was promoted.
Mark contacted the University's internal publication 'Bulletin' to arrange promotion of the survey in the next issue. Contact was made with Hampshire County Council's on-line services to broadcast the survey over HantsWeb/Hantsnet. On Monday both of these forms of promotion came into being.

FUEL CRISIS SURVEY

HAVE YOUR SAY

TRG
Transportation
Research Group

Drivers' Survey

How did you and your household cope last week and what changes did you make to your daily routine and journeys?
We hope this survey will help Government and other bodies take stock of how we use our cars and what this means for their policies.

Return this questionnaire by 29 September and you could win £250.

For this survey to be a success it is important that you try to remember as much as you can from the last week (week commencing Monday 11 September).

Queries? Telephone 023 8059 2834 or email M.E.Beecroft@soton.ac.uk.

SECTION 1 - ABOUT YOU

Age (in years): _____ Sex (male/female): ____ Postcode: ____________

Main occupation (please be as specific as possible): ____

Please list the details of all other people in your household:

Age	Sex (male/female)	Occupation	Relationship to you

Which of the following best describes where you live (please tick one only):
☐ remote area ☐ village ☐ town ☐ city (suburban) ☐ city (central/inner area)

How many cars or vans does your household have? (please include company owned vehicles): ____

Please describe the year, make, model and size of the vehicle you use the most for driving (e.g. 1993 Ford Escort 1.8 Diesel): ___________________

During a normal week how many miles do you drive using this vehicle? (please estimate to the nearest 10 miles): ___________

How often do you normally refuel this vehicle (please tick only one):
☐ at least twice a week ☐ weekly ☐ fortnightly ☐ less than every two weeks

At the beginning of last week did you manage to refuel or already have at least half a tank of fuel in -

- your vehicle (yes/no)?_____
- other household vehicles (if applicable) (yes/no)?_____

Do you have access to a bicycle (yes/no)?______

All information provided will be treated in the strictest confidence.

SECTION 2 - YOUR JOURNEY TO/FROM WORK

If you do not work each day please go to Section 4.

Please give us details of how you **USUALLY** travel to/from work:

How do you get there (please tick more than one if, for example, your journey involves using a bus and a train):

☐ car - drive alone ☐ train ☐ work from home

☐ car - drive with passenger(s) ☐ cycle ☐ other (please describe) ____________________

☐ car - as passenger ☐ walk

☐ bus ☐ taxi

How many miles is it from your home to work? ________

How many minutes does it take to get from home to work? ________

What time do you arrive at work (to the nearest ¼ hour, e.g. 8.45 am): ________

If you did not change your journey to/from work (as you described above) on any of the days last week because of the fuel crisis please explain why below.

If you did change your journey to/from work on one or more days last week because of the fuel crisis please provide as much explanation of how below - **we are interested in any changes to your usual journey to/from work as well as the affect these changes had on other journeys and aspects of your or your household's routine**. Please indicate (if applicable) how good or bad you found this change of experience to your journey to/from work.

____

____

____

____

____

____

____

____

____

____

____

If you did change your journey to/from work:

Have you done so before in the last six months in the way you describe above (yes/no)?____

Would you travel to/from work in this way again after the fuel crisis (yes/no)? ____
(please explain why or why not below).

____

____

All information provided will be treated in the strictest confidence.

SECTION 3 - YOUR JOURNEYS MADE WHILE YOU ARE AT WORK

If you do not work or do not make any journeys while you are at work please go to Section 4.

Please give us details of the journeys you **USUALLY** undertake for your work:

How often do you make long journeys at work, i.e. 5 miles or over (please tick one only):

☐ every day ☐ at least twice a week ☐ once a week ☐ at least once a month ☐ less than once a month

What form(s) of transport do you usually use for these journeys: ____________________

How often do you make short journeys at work, i.e. less than 5 miles (please tick one only):

☐ every day ☐ at least twice a week ☐ once a week ☐ at least once a month ☐ less than once a month

What forms(s) of transport do you usually use for these journeys: ____________________

What is the main purpose of these journeys (please describe)?: ____________________

If you did not make any changes to journeys undertaken at work on any of the days last week because of the fuel crisis please explain why below.

If you did make changes to journeys undertaken at work on one or more days last week because of the fuel crisis please provide as much explanation of how below - **we are interested in any changes to your usual journeys or routine while at work as well as the affect these changes had on other journeys and aspects of your or your household's routine**. Please indicate (if applicable) how good or bad you found this change of experience to the journeys you make at work.

If you did make any changes to journeys undertaken at work:

Have you made such changes before in the last six months in the way you describe above (yes/no)?____

Would you deal with work journeys in this way again once the fuel crisis is over (yes/no)? _____
(please explain why or why not below).

All information provided will be treated in the strictest confidence.

SECTION 4 - GETTING YOUR CHILDREN TO/FROM SCHOOL OR PRE-SCHOOL

If you do not have any children of (pre-)school age please go to Section 5.

Please give us details of how your child(ren) **USUALLY** get(s) to/from (pre-)school:

Who accompanies your child(ren) to/from (pre-)school (please tick the one that applies)?:

☐ no-one ☐ you ☐ your partner ☐ other (please describe) ________________

☐ child(ren) unaccompanied

How do(es) your child(ren) get to/from (pre-)school (please tick more than one if, for example, the journey involves using a bus and a train):

☐ car ☐ people carrier ☐ bus ☐ walk ☐ cycle

☐ minibus ☐ taxi ☐ train ☐ other (please describe) ________________

<u>If a car or people carrier is used</u>, please tick one of the following to indicate whether your child(ren):

☐ is/are taken alone ☐ is/are taken along with other parent's children

☐ is taken by another parent ☐ is/are taken in a car sharing arrangement

To the nearest mile, how far is it from your home to (pre-)school? ________

How many minutes does it take to get from home to (pre-)school? ________

If you have children that go to different (pre-)schools please indicate and explain anything not covered above:

____

____

If there were <u>no</u> changes to the journey to/from school on <u>any</u> of the days last week because of the fuel crisis please explain why below.

If you <u>did</u> make changes to the school journey on one or more days last week because of the fuel crisis please provide as much explanation of how below - **we are interested in <u>any</u> changes to the usual journey as well as the affect these changes had on other journeys and other aspects of your or your household's routine**. Please indicate (if applicable) how good or bad you found this change of experience in getting your child(ren) to/from school.

____

____

____

____

____

<u>If you did make any changes to the journey to/from school:</u>

Have you made such changes before in the last six months in the way you describe above (yes/no)?____

Would you deal with school journeys in this way again once the fuel crisis is over (yes/no)? ____

(please explain why or why not below).

____

____

All information provided will be treated in the strictest confidence. 4

SECTION 5 - GROCERY SHOPPING

Please give us details about how you **USUALLY** obtain your groceries:

Which person in your household usually buys the groceries? ____________

How often do you/they buy a large amount of groceries (three or more bags)?

☐ daily ☐ at least twice a week ☐ weekly ☐ fortnightly

☐ monthly ☐ less than once a month

How far do you/they travel to buy these (in miles): ____

What form(s) of transport do you/they use ________________

How often do you/they buy a small amount of groceries (one or two bags)?

☐ daily ☐ at least twice a week ☐ weekly ☐ fortnightly

☐ monthly ☐ less than once a month

How far do you/they travel to buy these (in miles): ____

What form(s) of transport do you/they use ________________

Which of the following statements is most relevant to the person who buys the groceries:

☐ Usually journeys made to buy groceries are only for that purpose.

☐ Usually buying groceries is combined with other journeys that are being make.

If your household did not change its routine of buying groceries at all last week because of the fuel crisis please explain why below.

If your household did change its routine of buying groceries because of the fuel crisis please provide as much explanation of how below - **we are interested in any changes to your grocery shopping as well as the affect these changes had on other journeys and aspects of your or your household's routine.** Please indicate (if applicable) how good or bad your household found this change of experience concerning grocery shopping.

If you changed your routine of buying groceries:

Have you made such changes before in the last six months in the way you describe above (yes/no)?____

Would you deal with grocery shopping in this way again once the fuel crisis is over (yes/no)? ____

(please explain why or why not below).

All information provided will be treated in the strictest confidence.

SECTION 6 - OTHER CHANGES TO YOUR HOUSEHOLD ROUTINE/TRAVEL

Please list other trips that you or your household usually makes on a regular basis that have been changed by the fuel crisis (e.g. library, pub, friends, pension, DIY, cinema, gym ...) – include purpose/frequency/ return distance (e.g. 'visit relative / weekly / 15 miles'):

1.________________ 4.________________

2.________________ 5.________________

3.________________ 6.________________

Please explain how they have been disrupted and how you have tried to cope:

As a result of the fuel crisis did you try to combine parts of your daily/weekly routine and travel during last week (either by combining trips together that you make yourself, combining trips you make with those of other members of your household or combining trips your household makes with those of people in other households) (yes/no)?__

If 'yes' please try and explain as clearly as possible how you combined trips or activities and indicate why you have not previously dealt with your daily routine in this way.

As a result of the fuel crisis did you cancel or postpone any activities or trips last week that you would normally have made (yes/no)?___

If 'yes' please try and describe the activities and trips concerned including what form of transport you would have used and the consequences of cancelling them.

*activities:*________________

*consequences:*________________

All information provided will be treated in the strictest confidence.

SECTION 6 - CONTINUED

Overall, which of the following best describes the affect the fuel crisis had on you during last week (please tick one only)?

☐ It caused me great difficulty
☐ It caused me some problems but not as bad as I expected
☐ It caused me no problems
☐ I found things better than usual

Have you made greater use of the phone/Internet during the fuel crisis (yes/no)?____

If 'yes' please explain how and why. If 'no' please explain why not.

Have you sought information concerning public transport because of the fuel crisis that you would not normally have done (yes/no)?______

If 'yes' please explain why and from where you sought information and if or how it helped you:

In view of your experiences last week, if on a continuous basis fuel was rationed which of the following would you be most likely to do (please tick only one):

☐ reduce the number of car trips you make ☐ buy a smaller/more fuel efficient vehicle
☐ move house to be nearer regular activities

All information provided will be treated in the strictest confidence.

SECTION 7 - YOUR VIEWS ABOUT THE FUEL CRISIS

Please indicate to what extent you agree with the following statements:

"The fuel crisis has made me realise that I use my car more than necessary during my normal life."
☐ strongly agree ☐ agree ☐ no opinion ☐ disagree ☐ strongly disagree

"The air pollution has been noticeably better during the fuel crisis."
☐ strongly agree ☐ agree ☐ no opinion ☐ disagree ☐ strongly disagree

"I have seen less traffic during the fuel crisis."
☐ strongly agree ☐ agree ☐ no opinion ☐ disagree ☐ strongly disagree

"I found the pace of life seemed to slow down."
☐ strongly agree ☐ agree ☐ no opinion ☐ disagree ☐ strongly disagree

"I found people were friendlier and more prepared to talk to one another."
☐ strongly agree ☐ agree ☐ no opinion ☐ disagree ☐ strongly disagree

"With less traffic on the roads I enjoyed driving more."
☐ strongly agree ☐ agree ☐ no opinion ☐ disagree ☐ strongly disagree

"I consciously drove more slowly than usual to conserve fuel."
☐ strongly agree ☐ agree ☐ no opinion ☐ disagree ☐ strongly disagree

"If fuel duty was to be reduced, it would be fairer to charge for using roads at times of heavy traffic."
☐ strongly agree ☐ agree ☐ no opinion ☐ disagree ☐ strongly disagree

"I support the government in not reducing fuel taxes as a result of the crisis."
☐ strongly agree ☐ agree ☐ no opinion ☐ disagree ☐ strongly disagree

"The fuel crisis has made me realise that we all depend a lot more on diesel oil and petrol than I realised."
☐ strongly agree ☐ agree ☐ no opinion ☐ disagree ☐ strongly disagree

"I believe that Government should bring in more policies aimed at cutting our dependence on diesel and petrol."
☐ strongly agree ☐ agree ☐ no opinion ☐ disagree ☐ strongly disagree

AND FINALLY.......P.T.O.

SECTION 8 - OTHER THINGS YOU WOULD LIKE TO SAY

We would be very pleased to learn of any other affects the fuel crisis has had on your daily routine and travel and what you would like to see happen in the future to the levels of car use, fuel tax and related matters.

THANK YOU FOR COMPLETING THE QUESTIONNAIRE AND SHARING YOUR VIEWS

If you want to be entered into the prize draw (see below) please give your name and address.

Name:.................................Address: ....

.... Postcode

THE PRIZE DRAW

Everyone who returns this form with their name and address will be entered for a prize draw to be made on 16 October 2000. The person named on the first randomly selected form will receive a cheque for £250. (subject to terms and conditions below).

Terms & Conditions of Draw: (1) Only those aged 18 and over on 29/09/00 will be eligible for the draw. (2) Only one entry per household. (3) All entries must reach the University of Southampton before 29 September 2000. (4) The Winner will be notified by post.

Please return you completed questionnaire (no stamp required) to:

Transportation Research Group
Dept. of Civil and Environmental Engineering
University of Southampton
FREEPOST LICENCE NO. SO 286
SOUTHAMPTON SO17 1YN

All information provided will be treated in the strictest confidence.

12 Crisis or Drama? – A Summary of the Response to the Fuel Crisis in the UK

GREG MARSDEN AND MARK BEECROFT

Introduction

This book has presented a range of analyses on the changes to personal and business practices that were made during the fuel shortages of September 2000. This chapter offers further insight through examination of additional information collected routinely by business, leisure, transport operators and local government. The chapter provides an overview at local, regional and national levels of the measured impacts of the fuel crisis, comparing the lifestyle and business impacts with the transport impacts. Methods of data collection included Internet searches, written requests for information and telephone interviews conducted by the authors who would like to register their appreciation to all the individuals and institutions who cooperated with their research.

The chapter assesses the extent to which routinely collected data sources can be used to understand the impacts of special events such as the fuel shortages. The more detailed investigations presented elsewhere in the book provide an important assistance to the interpretation of data presented in this chapter which itself seeks to perform the reciprocal function of validating the findings of the microscopic studies of behavioural change undertaken by other authors in the book.

The chapter presents a series of results of the effects of the fuel shortages on lifestyles, business and local government. The role of the media as a source of information and communication is discussed. Data are then presented on traffic levels and mode share. The chapter concludes by examining the policy and transport lessons that can be gleaned from this 'big picture' analysis of the effects of the fuel shortages. Through this analysis we answer the question posed by the title of the chapter – were the fuel shortages a crisis or a drama?

Lifestyle Impacts

In attempting to assess the impacts of the fuel crisis upon the lifestyles of the UK public the most striking feature is the high degree of variability of responses. Nowhere was this better emphasised than in the press coverage. There were quirky tales such as that of the touring rock band from New Zealand 'The Mutton Birds' calling upon its UK fans to bring petrol to its gigs (Ananova, 2000j) and the news that the Prince of Wales decided to cancel his engagements and a night out with Camilla Parker Bowles rather than accepting an offer of petrol from the police (Ananova, 2000a). More mundane stories such as the cancellation of weddings and overseas holidays (often with the knowledge that insurance policies offered no coverage under the unique circumstances), as well as social events all added to the sense of a curtailment of the nation's normal activities (Ananova, 2000c; Gascoigne, 2000). However, equally in evidence were stories reflecting the nation's traditional stoicism like the news that a record 70,000 crowd attended the Leuchars Air Show causing the gates to be shut with disappointed spectators locked out on Saturday 16th September (Mceachran, 2000).

The crisis also demonstrated a capacity to impact upon more unexpected areas of life. In London a coroner investigating unusual deaths had to postpone all local inquests because pathologists, police officers, experts and witnesses could not attend the Hornsey coroner's court. Delays incurred would mean that examinations could not take place and bereaved families would be unable to make funeral arrangements (London, 2000a). A seemingly more positive impact from the crisis was that there were fewer cars on the road which meant that there were also fewer accidents. However, this was not good news for everyone as the insurance company Accident Care Group blamed a 62 per cent fall in profits for the quarter July-September 2000 on the same period in 1999 on the crisis (Kraft Burke, 2000).

Whilst accidents may have been reduced the police were kept occupied by localised reports of mini crime waves as members of the public strove to apply innovative and often illegal solutions to their personal petrol predicaments. Hounslow police reported a series of petrol thefts direct from cars and lorries (London, 2000b). Similarly in Enfield attempts were made to break into a petrol station and diesel was stolen from a storage facility at a golf club (London, 2000a).

What was clearly evidenced, both from the media coverage at the time and from commercial and academic communities with an involvement or interest in this field in the aftermath of the crisis, was the almost total absence of any attempt to study and assess the breadth and depth of impact of the fuel crisis upon lifestyles. It is possible that such organisations interpreted that the short lived nature of the fuel crisis meant that meaningful analysis of its impacts on lifestyle would prove unprofitable.

Given this paucity of research a most valuable source of information for this chapter has been datasets for the nation's leisure interests which are maintained on an all year round basis which serve to show, in a more detached and objective manner than that of sporadic media coverage, a range of impacts caused by the fuel crisis on the nation's lifestyles. However, it has to be acknowledged that these facts and figures were not designed, in terms of collection and presentation, to serve the purpose of assessing the impacts of the fuel crisis.

Sport

As a nation supposedly obsessed with sport, and particularly football, any disruption to this area of UK leisure interests would have been keenly felt. The crisis did not lead to the cancellation of any matches in the English and Scottish Football Premier Leagues and very few matches were cancelled in the lower leagues although many non-league and amateur matches were postponed. The English Zurich Premier Rugby programme also survived unscathed but all fixtures were cancelled in the English Professional Basketball League and many speedway and motocross events were postponed. Horserace meetings at Ayr, Catterick, Newbury and Newmarket went ahead on the 16th September but the Pontefract meeting on the 14th September was cancelled, as there would have been insufficient ambulance cover (BBC Sport, 2000; Basketball 365, 2000). However, the Newmarket meeting suffered from poor attendance according to the clerk of the course Nick Lees:

> We have been hit by two obstacles in not only the petrol shortage but the weather. So those things have conspired to make for a pretty disappointing crowd. I haven't got a figure, but it is low. (Ananova, 2000f)

The reference made above to the weather also points to an additional factor which must serve to qualify (although to what degree is it unquantifiable) any assessment of the impacts of the fuel crisis upon leisure and lifestyles. September 2000 was significantly wetter and colder than the same month in both 1999 and 1998 (Winpenny, 2000) and the extent to which such adverse weather conditions served to influence people's leisure activities could plausibly be significant.

The nature of the impacts of the fuel crisis on sporting activities was highly variable. Whilst the majority of professional sports continued with minor disruption it was the general UK public who were faced with more severe disruption in terms of cancellation, postponement and curtailment of their opportunities to both participate and view sporting activities. These distinctions between the impacts upon amateur and professional sports offer a clear example of how the media shaped public perception and received opinion of what happened during the fuel crisis. Whilst the media extensively reported developments relating to professional sport, attention to the impacts upon amateur sport was extremely patchy and coverage was generally restricted to local rather than national organs of the media.

Entertainment and the Media

If non-essential leisure trips were cancelled in the process of trip prioritisation during the crisis then this would be likely to be reflected in cinema attendance figures. This would be particularly so given the trends in recent years for smaller cinemas in urban centres to be closed and replaced by multiplex establishments on the edge of urban centres that are often considered only to be accessible by private transport. Unfortunately, the nature of the available statistical evidence renders any assessment of such arguments highly problematic. The statistics of UK cinema admissions (CAA/ACNielsen EDI, 2000) are collected on a monthly basis only. Therefore it is unclear as to what impact, if any, the fuel crisis week had upon the monthly total attendance figures. The figures reveal that September 2000 did see a substantial monthly fall in attendance compared to the previous two months and was followed by a significant rise in attendance the following month. However, precisely the same pattern of events occurred in 1999 leaving it impossible to argue with any justification that the figures were affected by the fuel crisis.

Similar difficulties are encountered when attempting to interpret radio listening figures (RAJAR/RSL, 2001). The total hours of adult radio listening were higher in the period of July-September 2000 than during any other quarter from March 1999 to March 2001. However, listening figures for Radio 5, the BBC's 24-hour News and Sport channel which provided up to the minute coverage of the crisis saw its second lowest figures in the period covered in this quarter. BBC and commercial local and regional radio stations saw an increase in listening figures during the quarter of the crisis but this increase was dwarfed by the growth in the following quarter. The fact that these statistics are only published in the form of quarterly summaries limits the extent to which the fuel crisis can be attributed any specific and measurable impact.

In contrast to the cinema and radio figures the television audience figures provide a much greater depth and breadth of information which facilitates a more developed assessment of the impact of the crisis. Table 12.1 (BARB, 2001) shows the average weekly viewing figures across all UK television for the month of September in 1998, 1999 and 2000. The most notable statistic is that the viewing figures increased by 1 hour 50 minutes during the week of the fuel crisis, a rise not seen elsewhere in the data since 1998. The audience level declined in early October before rising again towards the end of the month. (It is likely that this represents seasonal viewing patterns.)

Table 12.1 Television viewing figures

Week ending	Average weekly television viewing (hrs:mins) per person
06/09/1998	24:14
13/09/1998	24:05
20/09/1998	23:07
27/09/1998	23:08
04/10/1998	24:02
05/09/1999	23:25
12/09/1999	24:06
19/09/1999	24:30
26/09/1999	24:20
03/10/1999	25:09
03/09/2000	24:21
10/09/2000	24:05
17/09/2000	*25:55*
24/09/2000	25:56
01/10/2000	25:58

It might be thought that the beginning of the Sydney Olympics on the 13th September may account for some of the rise. However, the opening ceremony did not take place until Friday 15th September and hence the commencement of the vast majority of sporting action did not get underway until the week ending 24th September 2000. This is reflected in the viewing figures for individual programmes (BBC evening Olympic coverage in the week ending 24th September averaged around 6 million viewers, the opening ceremony on 15th September attracted only 2.5 million viewers).

The most dramatic statistical evidence that the UK public responded to the fuel crisis by watching the television is found in the viewing figures for news programmes. The figures for BBC1 news shown in Table 12.2 show substantial audience increases during the week of the fuel crisis from the Monday evening leading to peaks on the Tuesday and Wednesday evenings in which the audience was more than 50 per cent higher than at the same times on the previous week and equated to roughly a third of the total TV audience.

Table 12.2 Viewing figures (millions) for BBC1 news programmes which were amongst the 30 most watched BBC1 programmes of the week

News programme	Week ending 03/09/00	Week ending 10/09/00	*Week ending 17/09/00*	Week ending 24/09/00	Week ending 01/10/00
Monday 6pm	-	5.54	-	6.74	6.58
Monday 9pm	-	6.62	*7.52*	6.06	5.95
Tuesday 6pm	5.21	5.89	*7.72*	7.06	5.89
Tuesday 9pm	5.64	5.51	*8.41*	6.49	-
Wednesday 6pm	5.08	5.75	*7.44*	5.95	6.40
Wednesday 9pm	-	5.64	*8.57*	5.05	6.47
Thursday 6pm	5.90	5.71	*7.51*	-	-
Thursday 9pm	6.19	-	*7.71*	-	-
Friday 6pm	5.05	-	*6.81*	-	-
Friday 9pm	-	-	*6.28*	-	-

The audience figures for Sky News were even more dramatic. The percentage share of the total television audience for this channel has ranged from between 0.6 per cent and 0.7 per cent over the last two years. However, during the week of the fuel crisis this share tripled to 1.8 per cent. It seems likely that this can only be attributed to the fuel crisis as the channel devoted up-to-the-minute live coverage to the story and from the week after the crisis normal audience share was resumed. Nick Pollard, Head of Sky News commented:

> In many ways, this has been the perfect story for rolling news...We've not only covered the story, we've developed ways of giving lots of micro-information as well. People are interested in the developing story, but they also want to know what the implications are for them. We started a crisis information update on Tuesday and introduced rolling tickertape, which is the perfect vehicle for detailed information which you can keep repeating. (Cozens, 2000)

The channel BBC News 24 experienced a similar surge in viewing when a record 400,000 viewers tuned in on the Tuesday and Wednesday evenings of fuel crisis week just before the Six O'clock News, the previous high was 260,000 following the Concorde crash in Gonesse, France on 25th July 2000 (BBC News, 2000) and the average rating for the channel was around 20,000. Commenting on the overall interest in the news coverage, Steve Anderson, Head of News and Current Affairs at ITV stated:

> When a story breaks that touches the lives of everyone, affecting schools, hospitals and emergency services, people will tune in. Information is at a premium and a lot of people want it. (Cozens, 2000)

The demand for information on the crisis even led to the current affairs programme 'Tonight with Trevor McDonald' on ITV outscoring Channel 4's 'Big Brother' by 6.7 million viewers (37 per cent of the total TV audience) to 4.6 million viewers (21 per cent) (Cozens, 2000).

What cannot be gauged from this analysis of viewing habits are the reasons why certain programmes were watched. Did viewers watch programmes about the fuel crisis because of a personal need to know the state of the situation in order to plan their household activities and travel or were people watching because it was interesting?

Telecommunications

The public also took more proactive measures to obtain the latest information on the crisis by utilising telecommunications services. BT reported that over 434 million calls were made on their lines on Thursday 14th September, an increase of some 33 per cent on the previous Friday and NTL reported that traffic on their call network had increased by about a third (Ananova, 2000d; Vnunet, 2000). Part of these rises were due to a marked increase in calls to directory inquiries with many asking operators for the phone numbers of local garages. Similarly the Internet was used as a major resource to help drivers find petrol during the crisis. A series of unofficial sites claiming to offer up-to-the-minute information on availability were set up including: Fuel Today (www.fueltoday.co.uk) and Petrol Busters (www.petrolbusters.co.uk). Indeed, Petrol Busters had to strengthen its infrastructure to handle traffic of one million hits an hour (Ananova, 2000b; Vnunet, 2000).

Telecommunications services were not only used to obtain crisis information, they were also harnessed to enable business to function, providing virtual accessibility in the absence of physical accessibility. BT reported that bookings to use its teleconferencing service doubled from the usual 1,100 bookings a day to 2,200 with an average of five people to a call and in some cases twenty to thirty people on the line on Tuesday 12th September. On the 13th September some 2,700 conference calls were booked with a further 1,800 calls being made through web bookings or without a managed service. The number of new teleconferencing accounts being opened had increased five-fold during the crisis week. BT also stated that videoconferencing bookings increased by some 33 per cent during the crisis (Vnunet, 2000).

A random sample survey of 35 companies at the end of the fuel crisis week conducted by New Work Solutions revealed that around 70 per cent of companies used teleworking as a solution during the crisis. How many of these companies already sanctioned the use of teleworking prior to the crisis is not stated (BT, 2000). The approach of the companies to teleworking fell into three categories, although the proportion using each category of response was not stated:

- *Those extending flexible working to cover the crisis*- relaxed core hours requirement in flexi time schemes, instructed managers to allow flexible working, promoted existing policies to a wider group of staff.
- *Ad hoc teleworking on a temporary basis only*- firms with no experience/acceptance of the practice took it up in emergency conditions, recognised benefits of the practice in the emergency but many planned to revert to normal practice as soon as possible.
- *No home working*- did not explore alternatives, many insisted staff came to work or took holiday. Some clearly could not use teleworking and security was cited as a barrier for financial service firms. Many others were simply complacent and the crisis did not motivate an exploration of alternative practices.

Hospitality and Tourism

The implications for the hospitality industry from the disruption caused to business by the fuel crisis were also mixed. The Hotel Booking Agents Association reported a dramatic daily increase in both reservations and cancellations. Whilst provincial hotels were hardest hit by cancellations, demand for city centre accommodation by businesses seeking to ensure that key staff could get to work offset the cancellations caused by those unable to travel. Many conferences were cancelled and there was a marked downturn in advance bookings. Problems within the hospitality industry were compounded by the loss of staff due to access problems during the crisis. This meant that in some cases companies had to deal with 50 per cent more transactions with up to 40 per cent less staff (HBAA, 2000). However, within a week of the end of the crisis London hotels reported an upturn in reservations, and conference bookings were quickly back to normal (Caterer, 2000).

It was not possible to obtain data on visitor attendance or occupancy for a wide range of tourist services and facilities either because data was not kept, it was yet to be produced for the year 2000, or companies/institutions were unwilling to divulge the information for commercial reasons. Many of these organisations suggested that the impact of the fuel crisis had been significant at the time but was (fortunately) too short lived to require further research and analysis. The subsequent flooding problems, the long term disruption caused by the Hatfield Rail Crash, and the Foot and Mouth crisis impacted upon the tourism industry in such a prolonged and serious manner that the problems caused by the fuel crisis were dwarfed by comparison. Indeed, the Foot and Mouth crisis was having such a profound impact upon tourism at the time that the authors contacted institutions from the sector that research was ongoing and the prospect of future published analysis of the impacts was strong.

Some organisations were willing and able to provide data on the impact of the fuel crisis and this can at least give a flavour of the consequences for tourism. Table 12.3 (National Trust, 2001) shows the visitor attendance figures for all National Trust properties. The figures illustrate that attendance was indeed lower in September 2000 than in the same month in 1998 (-2.7 per cent) and 1999 (-3.2 per cent). However, the same could be said about the months of August (-2.1 per cent on 1998 and -2.3 per cent on 1999), October (-1.8 per cent on 1998 and -3 per cent on 1999) and November (-1.6 per cent on 1998 and -3 per cent on 1999)- indicating a general downturn in attendance in the autumn of 2000, perhaps also attributable to the flooding and the disruption to rail travel caused by the Hatfield crash.

Small differences in patronage levels should not however be taken to be insignificant differences. If the same percentage drop in patronage from August 1999 to August 2000 had been observed from September 1999 to September 2000, 79,000 extra visitors would have visited National Trust Properties. It is also important to note that the fuel crisis had a serious effect over a period of about seven days, only one quarter of the monthly reporting period.

Table 12.4 (Scottish Tourist Board, 2001) shows Scottish Tourist Board figures on visitor occupancy of all types of tourist accommodation for the period July-December 1999 and 2000. The table shows that the percentage drop between September 1999 and 2000 in both bed and room occupancy is greater than in any other month. The degree to which this statistic can be attributed to the fuel crisis must remain questionable. Nevertheless, the fact that the relatively dispersed and topographically inaccessible nature of much of Scotland renders dependence on private motorised transport for all people (including tourists) acute, compared to England, may serve to suggest a marked impact by the fuel crisis upon the Scottish tourist industry. However, unless comparable figures become available for the English and Welsh tourist industries this remains an unproven argument.

Table 12.3 National Trust property attendance figures

	Visitor Attendance Figure
1998	
August	7,884,644
September	9,118,713
October	10,056,946
November	10,188,237
1999	
August	7,905,110
September	9,160,908
October	10,182,258
November	10,335,305
2000	
August	7,725,434
September	*8,873,705*
October	9,877,272
November	10,030,161

Table 12.4 Scottish Tourist Board accommodation statistics

Month	Bed Occupancy (%)			Room Occupancy (%)		
	1999	2000	Change	1999	2000	Change
July	57	55	-2	71	69	-2
August	63	62	-1	77	76	-1
September	*56*	*53*	*-3*	*73*	*69*	*-4*
October	47	46	-1	62	63	+1
November	34	34	-	52	52	-
December	28	30	+2	38	42	+4

The fact that the months of October and November saw very little change in occupancy suggests that the flooding problems that followed in the wake of the fuel crisis had very little impact upon tourism in Scotland. This also suggests that it was unlikely that poor weather alone could account for the fall in occupancy figures during September 2000.

Coach tour operators represent an area of the tourist industry that is highly dependent upon transport and they inevitably suffered during the fuel crisis. The Coach Tourism Council, the Confederation of Passenger Transport and the British Inbound Tour Operators Association all gave evidence that coach operators lost business due to cancelled coach trips and they predicted that the affair would harm UK's reputation as a tourist destination and hit bookings in 2001 (Whitaker, 2000; Worcestershire, 2000).

Business

Shopping

If there is one activity which can rival sport in the affections of the UK public it is shopping. Consumer spending would be a key area in which to assess the impact of the fuel crisis upon lifestyles. The headline statistics provided a fairly positive indication. In its retail sales briefing for September 2000 the Office for National Statistics (ONS) reported growth in sales of 0.6 per cent compared to August 2000 and 4.6 per cent compared to September 1999 (National Statistics, 2000). An ONS statistician commented: 'There was no effect overall from the fuel crisis. Some stores said sales had dropped but others said they had risen'(Atkinson, 2000). Similarly, the British Retail Consortium's (BRC) September Retail Sales Monitor reported a growth in sales value in September 2000 of 1.7 per cent compared to August 2000 and 2.9 per cent compared to September 1999 (BRC, 2000a). The Royal Bank of Scotland's Scottish Retail Sales Monitor reported a growth in sales value in September 2000 of 5.5 per cent compared to September 1999 (BRC, 2000b).

The figures on retail performance were well above market predictions and led Neil Parker, economist at the Royal Bank of Scotland, to comment:

> The ONS said the fuel crisis had no effect on sales, which is counter-intuitive given all the anecdotal evidence. Perhaps we imagined the supermarket de-stocking and the impact of the fuel shortages. At face value, neither 6 per cent rates nor fuel blockades can stop us spending. (The Times, 2000)

An analysis of how the fuel crisis impacted on different areas of the retail sector provides mixed results. The BRC report reveals that out-of-town retailers suffered from reduced sales during the fuel crisis as customers reduced their car use. Furniture, DIY, gardening and electrical goods retailers all endured losses and department and other stores located in complexes like Bluewater, Lakeside and Trafford Park saw sales 'fall off a cliff' from the Tuesday of the fuel crisis. Mail order retailers also suffered reduced sales and delivery problems. Bill Moyes, Director General of the BRC offered his overview of the analysis contained in the BRC report:

> Non-food sales suffered during the week of the fuel crisis, with out-of-town stores being particularly hard hit. Food stores did benefit, however, as customers stocked up in anticipation of disrupted supplies. Overall we estimate that sales were more than £200 million lower than they would otherwise have been during that week. (BRC 2000a)

Evidence from individual retail companies supports this notion of a divide in the impact of the fuel crisis on the sector between food and non-food retailers. Table 12.5 details the impact of the fuel crisis upon non-food retailers as described by the companies themselves.

This 'catalogue of woes' provides a marked contrast to the performance of most food retailers during the fuel crisis as shown in Table 12.6. Whilst the John Lewis department stores suffered a 10.3 per cent drop in sales, the Waitrose supermarket arm of the business enjoyed a 24.9 per cent rise in sales on the same week in September 1999. This equated to an overall 6.6 per cent rise in sales for the company as a whole (Baker, 2000). The overwhelming cause for this increase in sales was seen by the media to be the result of panic buying of food and other essential goods, as consumers feared that supermarket supplies would run out as delivery services ran out of petrol. They also opted to stockpile goods as with their own petrol resources running low they could only justify, or practically manage, one trip to the supermarket.

This frenzied situation was not reported as wholly good news for the supermarkets. By the end of the week supermarkets were adopting rationing strategies for certain goods and they warned that the fragile supply chain was in danger of collapsing. Abattoirs warned that meat supplies could dwindle and livestock may have to be destroyed. Grampian Country Food Group warned that 4 million chickens would have to be slaughtered if fuel did not become available to transport food to feeding mills and the lives of thousands of pigs were also endangered.

Table 12.5 Non-food retailer impacts

Company	Type of retailer	Fuel crisis impact
JJB Sports	Sportswear	Comparable sales growth between August and mid-October down 2 per cent, The fuel shortage reduced profits by £1 million and turnover by £4 million (Phelps, 2000; Patten, 2000).
John Lewis	Department Store	Cost between £2-3.5 million in lost sales and sales were down 10.3 per cent on the same week in September 1999 (Kilgreen, 2000; Daily Express, 2000).
Marks and Spencer	Department Store	General sales down by an estimated 1.2 per cent (£21 million) for the 24 weeks to September 16th (Marks and Spencer, 2000).
New Look	Women's Fashion	First half of the financial year pre-tax profits down £2 million on 1999 due to 'fuel crisis and one-off costs'(Ananova, 2000k).
Oasis	Women's Fashion	Like-for-like sales down 20 per cent during the week of the fuel crisis (Ananova, 2000h; Felsted, 2000b).
Silentnight	Bed Manufacturer	Cost company £500,000 (The Independent, 2000b).
The Officers Club	Menswear	Cost company £250,000 (Menary, 2001).
Thorntons	Confectioner	Like-for-like sales for the 12 weeks to September 16th reduced by 1.1 per cent (Kilgreen, 2000).

The Federation of Bakers and the Dairy Industry Federation stated that demand for bread and milk would result in shortages and supply problems. In the face of widespread panic buying Asda's 240 stores nationwide doubled their usual quantity of fresh food and in some stores were limiting customers to three loaves of bread, two bags of sugar and six pints of milk. Sainsbury's introduced rationing in some stores and gave priority to delivery of bread, milk and baby food. All the supermarkets said on the 14th September that they had adequate stocks for 'the next few days', however, if the crisis continued the situation could change quickly (The Scotsman, 2000a; Elliott, 2000).

Large corporations dominate business headlines and have a duty to report trading gains and losses to shareholders. However, it would be wrong to assume that the full picture could be seen from analysing supermarkets or town centres. One of the options exercised by individuals during the fuel shortages was to shop locally. However, the gains to this part of the industry are not easily identifiable, except to note that the overall ONS data showed a rise in consumer spending which cannot be entirely attributable to panic buying in supermarkets.

Table 12.6 Food retailer impacts

Company	Type of retailer	Fuel crisis impact
Burger King	Restaurant	19 per cent fall in yearly operating profits blamed on problems of 16 drive thru outlets which suffered during the crisis (Investors Chronicle, 2001).
Inter Link Foods	Bakery	Turnover up 33 per cent and profits up 40 per cent in quarter encompassing crisis (Pain, 2000).
Kwik Save and Somerfield	Supermarkets	50 per cent increase in sales during fuel crisis week (Somerfield, 2000).
Sainsbury	Supermarket	Sold 150 per cent more produce than normal on 12th and 13th September (Elliott, 2000).
Tesco	Supermarket	5.1 per cent growth in like-for-like sales in half year encompassing crisis (Menary, 2000).
Waitrose	Supermarket	Sales increased by 24.9 per cent during crisis week (The Scotsman, 2000b).
William Morrison	Supermarket	5 per cent growth in like-for-like sales and 9 per cent increase in yearly pre-tax profits (Ananova, 2000e).

On-Line Retail

The consequences of the crisis for food retailers were not limited to their retail outlets. It also had an impact upon the online grocery delivery market. In the case of one retailer, Waitrose, the message on the impact was mixed or perhaps even confused. The Financial Times reported on 16th September that new customers to Waitrose@work, its online deliveries-to-offices service quadrupled during the week of the crisis (Grande, 2000). However, the Daily Telegraph (Lambeth, 2000) had reported on the 14th September that the service was being forced to stop operation due to fuel shortages, along with the Wine Direct service. It also reported that the Waitrose.com online grocery delivery service had been scaled down for the same reasons. Iceland had 20 per cent of delivery vans off the road for one day during the crisis and was reported along with tesco.com as only having enough fuel to last until the 18th September. The latter also set up a 'fuel crisis group' to look at ways to maintain its fuel supply for home deliveries (Lambeth, 2000).

Other online retailers also suffered as a result of the fuel crisis. It was perhaps ironic that even companies that did business by means of virtual access were still dependent upon traditional means of delivery such as the van. The London based online delivery service urbanfetch.co.uk, which promised deliveries within an hour ran out of petrol completely and arranged for its couriers to deliver by bicycle. Online car selling companies were particularly hit by the crisis, Tins.co.uk reported orders were down by 50 per cent compared to normal during the crisis week and as a consequence of the crisis it had an order backlog of over 150 undelivered cars. Jamjar.com suffered similar problems with 40 orders affected. The disruption of deliveries from motor manufactures caused by port blockades and the promise of providing each new car with £10 of petrol in the tank all served to hamper business (The Independent, 2000a).

Industry: Logistics, Distribution and Manufacturing

The problems which hit online car sellers during the crisis were not isolated. They affected the entire motor manufacturing and retail industry. The Society of Motor Manufacturers and Traders (SMMT, 2000) reported that restructuring work at several UK factories as well as the impacts of the fuel crisis listed above served to significantly impact on the industry in September and October 2000. Total car production in September 2000 was down 24.3 per cent on September 1999 and commercial car production was down 29.1 per cent. New car registrations in September 2000 were also down 8.9 per cent on September 1999. The sheer scale of the impact on the industry meant that very little recovery had been made by October 2000 as total car production dropped 16.1 per cent on October 1999 and commercial car production was down 15.9 per cent. It was not only car manufacturers who suffered from the distribution problems caused by the crisis. Utility International, manufacturer of trailers, was forced to place 60 orders on hold, close its factory and lay off workers in response to the fuel shortage (North East, 2000).

As has been illustrated by the problems of the motor industry, it was in the area of logistics and distribution that the fuel crisis had its most profound impact upon UK industry. As Financial Times reporter Kevin Brown stated: 'The fuel crisis has exposed how dependent British industry has become on "just-in-time" systems for distribution of goods'(Brown, 2000). In an effort to cut costs and improve efficiency companies no longer keep warehouses full of supplies but draw stocks from suppliers at short notice and this 'lean' approach to manufacturing proved highly vulnerable to disruption of the distribution process. This system of approach has been widely adopted with the consequence that logistics and distribution processes have become increasingly complex and integrated. As Richard Wilding, Lecturer in supply chain management at Cranfield University stated; 'We are not talking about a supply chain here, we are talking about a supply network. What is happening is that failures in one supply chain are quickly causing failures in others' (Brown, 2000). In response to the crisis, TNT Express placed orders to increase its stock of fuel by 20 per cent within a month of the fuel crisis, in preparation for any future problems whilst Securicor stated that its distribution division profits for September had been reduced by £2 million because of the crisis (Felsted, 2000a) and the haulier Eddie Stobart stated that his business had lost £600,000 during the crisis week (Doran, 2000). How much of this cost was due to the fact that many of his truckers suspended normal work activities to participate in protest convoys was not stated.

The headline figures of the impact of the fuel crisis upon manufacturing were, as in the case of the retail sector, by no means as dramatic as the effects were upon individual companies. The Chartered Institute of Purchasing and Supply provided what was seen as the key measure of the effect of the crisis upon manufacturing through its Purchasing Managers index, which is based on a range of questions to manufacturers. The index stood at 51.6 in September 2000 down from 51.8 in August 2000, however, any figure above 50 points represents growth (Denny, 2000; Ananova, 2000g).

While fuel shortages did not impact on overall output, the delays and disruption caused to deliveries meant supplier's delivery times were the worst since May 1995 and this indicated that the full effects were still being felt. This was evidenced in the index figure of 49.8 in October 2000 although by November 2000 recovery was underway with an index figure of 51.3 (Ananova, 2000l). Further details of the impact of the fuel crisis on haulage firms are provided in Chapter 10.

UK PLC

Both during and after the fuel crisis there was a great deal of speculation in the media regarding how the crisis would affect the UK economy. There were doomsday headlines drawn from renowned experts (such as Digby Jones, director-general of the CBI (Doran, 2000; Griffin, 2000; Pook, 2000)) on the economic cost of the affair which ran from millions to billions of pounds. When we look beyond the sensational headlines the picture is both more mixed and more complex, again reflecting the high degree of variability with which the fuel crisis impacted on UK life.

The Institute of Directors (IoD) issued a press release on 15th September projecting the likely cost to the UK economy of the fuel crisis (IoD, 2000). The IoD calculated that *if* the crisis continued for 2 to 3 weeks and the disruption to the economy was about 5 per cent of total activity for the first week (an estimate based on a straw poll of IoD members) rapidly declining thereafter, the loss in output could be around £1 billion (about 0.1 per cent of GDP at market prices). This was equivalent to around £40 per household. Perhaps unsurprisingly in the midst of the crisis the media leapt upon these figures stating that the crisis was wreaking destruction on businesses and crippling the economy. In the aftermath of the event, only the *Financial Times* reported the IoD's revised and greatly reduced cost predictions which was based on the fact that the qualifying criteria of their estimate, particularly in relation to the longevity of the crisis, were not met (Turner, 2000).

There was no doubt that the crisis had severe consequences for individual businesses, as evidenced by the British Chamber of Commerce's submission on the fuel crisis which detailed in great depth the impact felt by individual companies as reported to local chambers of commerce (BCC, 2000). Tables 12.7 and 12.8 give a summary of some of the impacts as reported by the St Helens and Peterborough Chambers of Commerce who conducted surveys of their members receiving 310 and 193 responses respectively.

In presenting its evidence the BCC recognised that the impact of the fuel crisis on UK business was more varied and complex than can be expressed in such figures. Sales and orders were lost, but many were simply taken by competitors or deferred to a different time period and so were not lost to the UK economy. A diverse range of costs were incurred by businesses. Some had to employ more staff to manage crisis conditions and provide information to customers. Efforts to catch up on lost output resulted in overtime payments. It was likely that the launch of advertising and marketing campaigns coinciding with the crisis caused both objectives and revenue to be lost. It should also be considered that the financial viability of UK businesses was affected by the crisis as it may have served to discourage both domestic and overseas investment in the UK economy.

Table 12.7 St Helens Chamber of Commerce survey

Impact	Percentage of businesses that said they were affected
Lost orders	28
Closure	6
Cut staff	6
Annual profit and cashflow problems	56
Expected longer term impact on sales	33
Expected relationships with customers and suppliers damaged	33

Table 12.8 Peterborough Chamber of Commerce survey

Impact	Percentage of businesses that said they were affected
Generally felt they were affected	86
Inbound deliveries delayed	59
Outbound deliveries delayed	56
Output decreased	26
Staff absences	22
Export difficulties	5
Lost sales	38
Financial difficulties	6
Temporary cessation	16

It would be expected that consumer and business confidence would have been adversely affected by the fuel crisis. A survey of 2,000 people in the UK undertaken between the 5th and 14th of September for the European Commission by the market research company Gfk certainly bore witness to this assumption (Crooks, 2000). When asked their views on their finances and the health of the economy the response equated to a fall in the aggregate index of consumer confidence from plus 1 in August 2000 to minus 5 in September 2000. This served to indicate that consumer confidence was as low as it had been since the global financial crisis of 1998.

However, as has been the case with a significant proportion of the evidence gathered on the impact of the fuel crisis, an alternative perspective can be presented. Business information group Dun and Bradstreet conducted a UK Business Expectations Survey in October 2000 in which 1,000 managers and finance directors were questioned on their views of the prospects for business in the final quarter of the year (Dun and Bradstreet, 2000). The survey revealed that 53 per cent of those questioned expected to increase profits, the same proportion as for the previous quarter. Of respondents 42 per cent expected increases in new orders, which was up 15 per cent from the previous quarter, but the proportion expecting increased sales dropped from 66 per cent to 63 per cent. This news was widely reported in the media as showing that the fuel crisis failed to dent business confidence (Ananova, 2000i, Wright, 2000, Mid Sussex, 2000).

Local Authority Services

Local authorities provide a wide range of services from education, through social services to refuse collection. Large numbers of people travel every day on Local Authority business and these movements were clearly affected by the fuel crisis. All Local Authorities have an emergency strategy for winter weather, where many staff may be unable to reach their workplace for short periods and these were used and adapted to good effect during the fuel shortages. One option that was exercised by many Local Authorities was to reduce the number of services operated to core services only. Such measures were supported by information dissemination, often through websites and the local media. Table 12.9 shows the types of services that were deemed non-essential and subject to postponement.

Local Authorities placed a priority on providing fuel for buses running school services and no examples of cancelled bus services were found. School meal services were also supported. Whilst providing transport to get children to school is within the control of the local authorities, provisions for teachers are not managed. Bungay High School near Norwich issued a letter to parents stating 'We are able to man the school because we have enough teachers who live within walking distance.' The

National Grid for Learning issued advice to schools that informed non-attendance due to fuel shortages should be treated as authorised absences.

Table 12.9 Local Authority services subject to disruption

Services cancelled or affected	Local Authority
Recycling collection service	East Hampshire, Wealden
Composter sales	East Hampshire
Non-emergency pest control visits	East Hampshire
Planning site visits	East Hampshire
Reduced home visits by housing services	East Hampshire
Reduced street cleaning and grass cutting services	East Hampshire, Lancashire
Community Transport Services – restricted operation	Lancashire, Dudley
Subsidised routes evening and Sunday services suspended	Lancashire
Day care transport running at 20 per cent	Lancashire
Normal road maintenance work suspended	Lancashire
Trading standards restricted operation to petroleum storage assessments	Lancashire
Unavoidable schools closure procedure put in place (winter weather routine)	Lancashire

A number of schools were forced to close due to staff shortages. Details on school closures are held at a local level and no national analysis has been conducted. A telephone survey conducted by the authors of seven local education departments, concentrating on less dense areas such as Cornwall, was unable to identify any records of school closures. However, South Norfolk Council (2000) noted about their own staffing levels 'Where schools were closed this accelerated staff problems as they were unable to obtain child minders at short notice.'

Local Authorities also reacted to the crisis by improving the efficiency of work related travel. Bath and North East Somerset Council encouraged staff to work from the nearest office base and to car-share. They also cancelled all 'non-essential meetings'. Had the crisis continued, South Norfolk Council noted that only a skeleton service would have been available at the main Council house run by staff who were able to attend work without the use of vehicles.

Fuel Diversity

South Norfolk Council had practical experience of the benefits of alternative fuelled vehicles during the fuel shortages. The Council has 11 Liquid Petroleum Gas (LPG) vehicles split between refuse collection, building, grounds maintenance and sewage maintenance. Had the crisis continued then these would have been the only vehicles available for service. Other Local Authorities with large fleets of electric vehicles (e.g. Hull, Coventry and Oxford) were able to continue operations of such vehicles unaffected by the fuel shortages.

Joynson Bruvvers Ltd, an office supplies delivery service have 3 LPG delivery vans - 2 Mercedes 314 Sprinter and a Vauxhall Combo in a fleet of five vehicles. These were able to run their service in Oxfordshire without problems (multi-drop - 40 drops per day). They have an LPG tank on-site and had 2 weeks supply of fuel, despite the diesel trucks which deliver LPG being unavailable (Davison, 2001).

Safeway also operate a fleet of natural gas trucks, which they were able to operate without problem during the crisis and contributed to the emergency plans they have in place, to ensure delivery to stores.

Whilst the situation surrounding the fuel crisis was somewhat unique, the longer term security of energy supplies is a major issue across Europe and the subject of a recent Green Paper (EC, 2001). The experience of fleets with some alternative fuelled vehicles during the crisis suggests that there are business planning benefits with alternative fuelled vehicles as well as environmental benefits and potential cost savings. Indeed, South Norfolk Council are examining expanding their purchase policy for LPG vehicles. Whiteing et al. (2002) in Chapter 10 also found evidence of an increased willingness to use and consider alternative fuelled vehicles within freight and logistics operators.

Traffic Flows

The Government and a number of Local Authorities have conducted analysis of the changes to vehicular flows and other associated impacts such as parking levels, accidents and the environment over the period of the crisis in a similar manner to the work of Eves et al. (2002) in Chapter 6. This section reports on and interprets these statistics. Traffic flow levels have a readily identifiable repeated pattern over the course of a week. However, daily variations and variations by hour of the day due to factors such as the weather can make it difficult to separate out the effects of special events from the noise of typical variation. Despite these caveats, a number of clearly significant reductions in traffic flow have been identified around the UK.

The DETR (now DTLR) has an automatic traffic count and classification network covering 135 sites across the UK. Following the crisis, a brief analysis of the changes in traffic flows was conducted (Hathaway, 2001). The analysis compared car and goods vehicle levels with the average flow levels from the same day of the week in August and up to the 10th of September when the effects of the fuel shortages began to take effect. A summary chart of the impacts is shown in Figure 12.1.

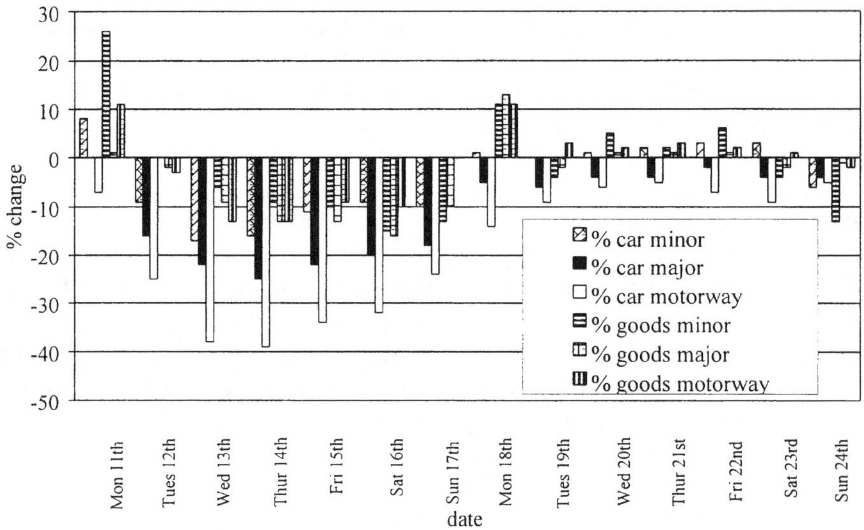

Figure 12.1 DETR measured effects of the fuel protest on road traffic (From data in Hathaway, 2001)

The reductions in flows are the average of several sites in each case and it is not clear what proportion of traffic is goods vehicle on each road type (although DETR suggest that changes to goods vehicle movements on minor roads be treated with caution due to their small numbers). Chapter 6 estimated that goods vehicles make up between 21 per cent and 23 per cent of traffic on the M27 and M6. Taking the average of these two values, then the total reduction in motorway traffic flows over the Monday to Friday of the fuel crisis week were -3, -20, -32, -33 and -29 per cent respectively, data which are consistent with that of Chapter 6. The data shows that goods traffic was less affected by the crisis than private vehicles. Indeed, on Monday 18th goods traffic increased by 10 per cent over a typical Monday on all roads – probably to clear the backlog of work.

Greater Manchester Transportation Unit (2000) compared flow data from the fuel crisis week with that of a comparable week in March 2000. At the height of the crisis traffic flow levels were 17 per cent to 33 per cent lower than March levels on the 9 'A' road sites assessed and 39 per cent lower on the M60. Typical average 7 day average flows (including the weekend of the 16th/17th) were between 11 and 20 per cent lower. Much lower drops in flow were seen on a road accessing the World Cargo Centre with typically 30 per cent goods vehicles. The maximum reduction on this road was 13.4 per cent and the 7-day average 7.4 per cent.

Leeds City Council (2000) conducted a similar analysis but comparing data to the week beginning 4th September. 24 hour flow totals dropped by 24 per cent with PM peak flows found to drop by 24 per cent. The corresponding figure for the AM peak was however only 16 per cent. Leeds City Council attributes these differences to 'a degree of prioritisation of journeys by drivers, with the journey to work being rated as the highest priority.'

Lancashire County Council (2000), in common with the other data presented found that the maximum flow reductions occurred on Thursday 14th September. They estimated maximum and average flow reductions from automatic count sites in the town centre (23 per cent and 13.5 per cent), outskirts of the urban area (32 per cent and 23 per cent) and on motorways (45 per cent and 34 per cent). The data from the fuel crisis week was compared with that of the previous week.

Hertfordshire County Council (2000) examined the changes in flows at 8 sites compared with the equivalent week in June. A control test in 1999 showed that variations in flow levels are of the order of 1 per cent. The average flow reductions were found to be between 11 per cent and 22 per cent for the Monday to Friday of the fuel crisis week. Peak periods showed drops of 9 per cent (am) and 13 per cent (pm).

The pattern of flow reductions is more complex than can be obtained through the daily total analysis in the official DETR Transport Trends report. Two of the four local authority sites examined the flow changes through the day and found the morning peak period flows reduced less than the evening peak flows. This perhaps represents the importance of work journeys and school trips. Indeed, data from Chapter 7 (Chatterjee and Lyons, 2002) shows that a lot of leisure, business and shopping car travel was not undertaken. The data all support the belief that longer journeys were sacrificed to enable more 'essential' local trips to continue. However, significant reductions in traffic levels on smaller roads, of the order of 14 to 20 per cent were still seen.

Parking

Leeds City Council also examined changes to parking levels in their public car parks. The study (Leeds City Council, 2000) notes that the percentage reductions in car park occupancy levels were far in excess of the reductions in traffic flow with 40 per cent reductions being observed for long stay car parks. The Council states that the most likely explanation for this is that trips to the city centre are more easily made by alternative modes than trips through the city centre. It is not however possible to say whether the drop in occupancy was a result of people choosing shopping centres closer to home, postponing trips or changing modes. It seems likely from the mixture of responses reported in Chapter 7 that it is a combination of all three.

Car Occupancies

Greater Manchester Transportation Unit (2000) also measured car occupancies at four sites on Friday 15th and Friday 22nd of September and compared the results with a previous survey from May 2000. Friday 22nd of September was European Car Free Day. The results are shown in Table 12.10.

Table 12.10 Car occupancy observations from Greater Manchester

Time period	Average number of occupants per car		
	17/5/2000	15/9/2000	22/9/2000
0800-0900	1.18	1.24	1.17
1000-1200	1.33	1.32	1.29

The data shows that European Car Free Day had no positive impact on vehicle occupancy levels compared with a normal day. However, the fuel crisis showed a small but significant increase in vehicle occupancy rates in the morning peak. The increase in occupancy means that 'for every 100 cars on the road normally, only 95 were use on 15th September 2000 to carry the same number of occupants.' (GMTU, 2000, p9). Chatterjee and Lyons (2002) found that 12 per cent of car using commuters had car-shared as a response to the fuel crisis.

Such a reduction in traffic levels without reducing the number of person miles seems an extremely cost-effective measure to pursue, given that there was no co-ordinated push for car sharing within the area and many of these agreements will have been made informally. It is also worth noting that the peak of the crisis was on Wednesday 13th and Thursday 14th and the figures on these days may have been even higher. It is clear from the rapid return to normal occupancy levels that the car-share response to the fuel shortages is not sustainable for longer periods, at least without employer support.

Accident Statistics

Most road accidents result from a combination of factors including driver inattention, inappropriate speed and poor visibility. The human error element of any accident makes the pattern of occurrence difficult to assess, particularly over short periods of time. The Police log all injury accidents in the UK and Local Authorities use the data to determine clusters of accidents or 'accident blackspots'.

A review of speed policy by the Government (DETR, 2000) has found that, 'In any given situation, the faster the average traffic speed, the more collisions there are'. Taylor et al. (2000) also found that accident frequency rises disproportionately with increasing speed estimating that a 10 per cent increase in mean speeds on urban roads can lead to a 21 per cent increase in collisions.

Intuitively, given the drop in flows on all roads observed during the fuel crisis, combined with anecdotal evidence and advice to adopt a more conservative driving style, one would expect accident levels to fall. Three Local Authority bodies attempted to quantify this change (Greater Manchester, Lancashire County Council and Leeds City Council). The results of the analysis are presented below in Table 12.11.

It is wise to treat the figures in the table as indicative as they depend on the reference levels against which they have been compared and are also subject to the randomness of accident occurrence as a data set themselves. The data sets for the fuel crisis week have been compared to the average values of the accident statistics from the equivalent week in the previous three to five years.

The data shows the difficulties in assessing the accident impacts of reduced traffic levels and perhaps speeds. The number of fatal and serious injuries in a typical week is small and the percentage changes are therefore more easily influenced by what could be freak events. However, the total number of accidents in any week is much higher and the direction of change in accident levels is therefore less subject to the random variation. The statistics show a fall in total accident rates of between 17 per cent and 45 per cent for the whole week (Sunday to Sunday).

Table 12.11 Accident statistics from 3 areas

	Fatal	Serious	Slight	Total
West Yorkshire				
1997	2	30	236	268
1998	2	13	269	284
1999	0	31	233	264
2000	2 (+50%)	23 (-7%)	178 (-28%)	203 (-25%)
Lancashire				
1995	2	27	77	106
1996	3	20	99	122
1997	1	14	88	103
1998	1	27	109	137
1999	4	20	86	110
2000	0 (-100%)	10 (-53%)	54 (-41%)	64 (-45%)
Greater Manchester				
	Killed and Seriously Injured			Total
1995	32			263
1996	16			226
1997	18			228
1998	18			285
1999	18			219
2000	24 (+20%)			202 (-17%)

These results are extremely encouraging in that they provide extra evidence that measures to reduce traffic levels will lead to a reduction in accidents. Whilst this data only covers three areas, all in the North of England, it should be possible for other local authorities to extract this data from the historical records and to provide a broader picture with greater confidence levels.

Air Quality

The work of Eves et al. (2002) in Chapter 6 described how changes to flow levels and speed profiles can affect toxic exhaust emissions and fuel consumption. These findings are supported through other research (De Vlieger et al. (2000)), which reports that fuel consumption in busy urban conditions can be 20 to 45 per cent higher than in non-congested traffic.

However, there are a complex series of interactions that occur between what is emitted at the tailpipe of a vehicle exhaust and how this affects local air quality. Local air quality is influenced by pollution imported from other areas, sources other than traffic, local topographic conditions and meteorological conditions (Marsden and Bell, 2001 ; McRae et al., 2000). The link between reduced traffic levels and emissions and improved air quality is not therefore straightforward.

A study was conducted for the DETR and the Environment Agency to investigate the air quality impacts of the reduction in traffic and associated emissions during the fuel crisis. The study used the UK Meteorological Office NAME model (Ryall et al., 1998) and compared the pollutant concentrations that would have been obtained with the usual flow profile with those concentrations measured at six urban background high precision (AUN) monitoring stations. Any differences between the modelled and measured concentrations would represent the pollutant concentration effect of the fuel crisis. The results of this study were however inconclusive for two reasons:

- the uncertainties in the modelling were too large; and
- there was no observed effect (Loader et al., 2001).

Large decreases in concentrations were seen over the weekend following the fuel crisis at all six sites. However, these reductions were also found at the Belfast Centre monitoring site which was unaffected by the fuel shortages. The drop in pollutant levels was therefore attributed primarily to meteorological conditions. The air quality in September 2000 was generally good, for reasons independent of the fuel shortage.

Following this study, the Environment Agency commissioned AEA Technology to conduct a further study to investigate the impacts of the fuel crisis on a wider range of sites and to examine the changes in air quality alongside the changes in traffic flow levels. This section summarises the key results from that report (Loader et al., 2001).

The study examined flow statistics at twelve sites. Four sites were selected to represent industrial areas, four to cover urban areas and four roadside sites. Nitrogen Oxides (NO_x), Particulate matter (PM_{10}) and Carbon Monoxide (CO - at roadside only) were examined. The proportion of pollutants attributable to transport increase from industrial to roadside sites. Data from September 1997, 1998 and 1999 was used to establish a reference case pollutant level against which to compare the results for September 2000. Estimates of the percentage reduction in pollutant levels were made based on the reductions in traffic flow and data from the National Atmospheric Emissions Inventory (Salway et al., 2000). The estimated percentage reductions with respect to the reference case were then compared with the measured reductions. The study concluded that:

> The estimated reductions in ambient concentrations of NO_x, PM_{10} and CO caused by this (reduction in traffic) were in *some* cases large enough to theoretically be detected by measurement (particularly near motorway sites).
>
> Ambient concentrations of NO_x, PM_{10} and CO did show a reduction, relative to typical average September values. However, in most cases, the reductions observed were greater than predicted by the measured traffic reductions alone. This is consistent with ambient concentrations being reduced by a combination of reduced traffic emissions and meteorological conditions.
>
> ...in most cases the impact on ambient concentrations of these pollutants appears to have been too small to clearly identify above the larger fluctuations caused by meteorology. (Loader et al., 2001)

Whilst there is some evidence to suggest that air quality was better than it would otherwise have been during the period of the fuel crisis, the data is not as conclusive as one might have thought from work examining solely emissions impacts. Clearly, climatic conditions are a dominant factor in determining air quality levels and it is also difficult to draw out definitive answers from short-term datasets. The implications for air quality management areas are somewhat more difficult to extract. This work shows that only small gains will be found when the climatic conditions are favourable for dispersion and measures to reduce traffic will have little effect. However, such measures could have a more significant impact when pollutants are not dispersed, arguably when the reductions are most required. It is also questionable however, whether many air quality management areas will provide the same level of traffic and therefore emissions reductions as seen during the crisis.

Mode Shift

Comparatively little data has been published on the effects of the fuel crisis on other transport modes. This may, in part, be a result of the extra organisational effort that public transport companies needed to undertake to ensure that fuel was available for their fleets and that drivers were available to man them. Those results which are available are presented by mode below.

Buses

The Home Office of the Government placed bus and coach operators ninth on a priority list for fuel supplies (shown in Appendix A). Most companies were able to maintain fuel supplies and normal services throughout the crisis. Where fuel shortages hit, bus companies reacted in a similar way to local authorities and cancelled non-essential leisure services or reduced service levels to a Sunday service equivalent. Blackpool Transport Services for example received no normal deliveries and operated only a shadow service. However, Blackpool was able to react and increase the frequency of tram services (Coach and Bus Week, 2000). Armchair Buses in Brentford had to cease the operation of two double decker buses for four hours through lack of fuel (TRANSIT, 2000).

The larger bus companies were able to act more strategically with Stagecoach hiring tankers to transfer fuel between operating companies to keep vehicles on the road. Fuel was transferred from Sunderland to Manchester by Stagecoach and FirstGroup claimed that no services were lost in Scotland or the South West (TRANSIT, 2000). Had the shortages continued for two or three more days, the shortages would have had a greater effect on service levels.

Data on patronage changes is held at a local level. Table 12.12 summarises estimated changes in patronage from the articles in TRANSIT and Coach and Bus Week referenced above.

Table 12.12 Changes in bus patronage

Company	Area	Change in patronage (%)
FirstBus	Glasgow	+4
Travel West Midlands	Birmingham	+4
Travel West Midlands	Around Birmingham	+5
Southern Vectis	Isle of Wight	+19

Whilst Travel West Midlands estimated that some of their increased patronage was a result of people who 'I'm sure haven't used the bus in years', Southern Vectis (2000) conducted the only analysis found on the make-up of the new patronage and their opinions. The Southern Vectis survey interviewed 653 people of which 106 were using the service as a result of the fuel crisis. Over 80 per cent of the new passengers felt that the bus service was a least as good as they had expected with nearly one half of these people believing the service to be better than expected. Principal complaints were high fares and bumpy roads.

It is difficult to put these changes in context, as detailed information regarding the total number of trips is not available except for the Isle of Wight. The 19 per cent increase in patronage was estimated to be equivalent to a 1.5 per cent mode shift from car to bus (Southern Vectis, 2000). Changes of this order of magnitude imply that only a small (yet significant) proportion of those car trips that were not undertaken, as evidenced from the flow data, were substituted by bus travel.

Rail

The rail industry saw a more impressive surge in patronage over the period of the fuel crisis, which the Association of Train Operating Companies placed at around 20 per cent (ATOC, 2000). ATOC stated that a 20 per cent increase in patronage did not correspond to a 20 per cent increase in profit as the operating companies had to pay for extras such as accommodating workers in hotels so they could reach the station.

Like the bus industry most companies were unaffected by the shortage of fuel with very few examples of train services being terminated early. London-Paignton services were stopped at Exeter and Brighton-Reading services stopped at Basingstoke to conserve fuel. Anglia Railways cancelled some Crosslink services and GNER had to cancel one diesel service.

The National Rail Enquiry Service received more than 4 million calls between Monday 11th and Thursday 14th September. This compares to 5.2M taken over the previous one month period. This peaked during Thursday 12th September when more than 1.3M calls were received in the 12 hours between 6am and 6pm when the shortages were at their most acute. Extra phone lines and call staff were made available to ensure that the public continued to get through. This proved to be good practice for the events following the tragic rail crash at Hatfield in October, when the total calls eclipsed the fuel crisis record in reaching 10.5M for the month.

Part of the increase in patronage is most likely due to rail being the preferred solution for the longer distance car journeys that were cut in response to the fuel crisis – evidenced by the drop in motorway flows. However, TRANSIT magazine also reported on increased passenger numbers for local services. Unfortunately, no long-term mode shift was sustained due to the rail crisis that ensued in October following the Hatfield crash (House of Commons, 2000).

Other

The Manchester Metro light rail service saw significant increases in patronage during the fuel crisis period. Patronage levels were up by 40 per cent on Monday 11th September, 60 per cent on Tuesday, peaking at 104 per cent on Wednesday. Patronage levels were 65 per cent higher on Thursday and 40 per cent higher on Friday the 15th (Perkins, 2001). No extra passenger services were run to cope with the extra demand but a major public relations exercise was launched to encourage people to travel off-peak or earlier or later than they usually would during the peak periods to spread the load. Additional services were run later at night and in the morning in conjunction with the bus companies to ensure that workers from both modes were able to get to and from work. The StageCoach Supertram service in Sheffield also saw increased patronage. However, StageCoach were unable to increase service frequency during the peak periods as all of the trams are currently in use during the peaks (Shucker, 2001).

Domestic air travel was largely unaffected by the fuel crisis. Domestic air travel covers long trips such as Glasgow-London and such trips are unlikely to have been competing with car trips over the same distance. BMi reported no significant change to their patronage levels. This is probably because internal flights do not compete with car use and no transfer was likely. EasyJet were reported to have flown an empty jet from Luton to Liverpool to siphon off some fuel (Evans, 2000).

Conclusions

This book is entitled 'Transport Lessons from the Fuel Tax Protests of 2000' yet throughout the book, and even within this chapter, the events of the 11th to 15th of September are referred to as the fuel crisis. Whilst the evidence clearly points to the potential for the fuel shortages to have escalated to crisis levels had they continued, the evidence on lifestyle and business impacts of the shortages show mixed messages about the response. There appear to have been a number of winners as well as losers resulting from the fuel shortages – out of town shopping developments suffered during the period but local shops saw increased business, perhaps giving us an accelerated taste of Government policy! Supermarket sales boomed whilst Drive-Thru restaurant profits dropped.

Whilst there were some winners and losers in business, the fuel shortages did show how reliant we are on Just-in-Time deliveries and how fragile such a business process can be. Supply chain problems would have been much worse had the crisis continued one or two days more. However, whilst nobody would like to see businesses go under as a result of protest action, how sympathetic should we be to businesses that suffered profit losses as a result of the fuel shortages when the transport profession often bemoans the transport implications of these processes in better times?

Some businesses reported significant losses as a result of the fuel crisis with staggering estimates reported in annual shareholder reports. However, it is not clear to what extent these estimates are justified or a smokescreen for wider underperformance. It is clear from the media reporting, and our subsequent investigation of the basis for the often large losses estimated during the fuel crisis week, that artistic license was often used by business representatives and they found an eager audience in the media to report this.

There is ample evidence that people prioritised their activities during the crisis, cutting down on non-essential leisure activities. The figures do not however show the size of drop that one might have expected given the drop in traffic flows. This is partly as a result of the statistics often being available as monthly data which has been a problem in enabling us to identify the impacts of this relatively short event.

It seems clear from the data on the drops in vehicular flow and the much smaller increases in public transport patronage figures that people did not simply mode shift to complete the activities that they would have conducted by car. Instead, it seems that people either postponed these activities or cancelled them. The high level of use of National Rail Enquiries Service indicates that people were willing to find out about public transport options. The relatively modest increases in public transport usage suggest that they did not feel that they could conduct their current activities by other means. This is a clear illustration of how car dependent we are as a society and may perhaps temper our future expectations of mode shift?

The figures for mode shift to LRT were more encouraging than those for buses, which gives some indication of the superior perceived quality of LRT as a public transport mode. However, success was limited by a lack of 'on-tap' capacity to manage the rise in peak period patronage.

The reductions in travel provided the expected benefits in terms of reduced congestion and also reduced accidents. The impacts on air quality were however not so clear, which is perhaps surprising given the large drops in flows reported. Meteorological conditions were much more important than transport emissions during the period of the crisis. Further investigation of this phenomenon is warranted under less favourable weather conditions, where pollution does not disperse so easily. This will enable us to identify what sort of levels of flow reduction will be required to make a real and measurable difference to air quality (and not just emissions) in the recently declared Local Authority air quality management areas.

An interesting side effect of the fuel shortages appears to be the benefits that local authorities and some businesses derived from the use of alternative fuelled vehicles, which were not affected by the fuel shortage. There is some evidence that the fuel shortages have encouraged more organisations to look to the benefits of a more diverse fuel mix.

Local Authorities seem to have been better placed to manage the fuel crisis than Central Government with a more rapid and focused response. Many Local Authorities were able to implement an amended 'bad weather' strategy, usually employed during periods of heavy snow. Another element that stands out from this research is the reliance on people who can walk or cycle to work to provide 'emergency service' for those who were unable to attend. The survey by New Work Solutions highlighted two contrasting approaches – companies who encouraged home working and others who insisted that absences were taken as holiday. Whilst the second of these options might seem unfair on the face of it, it implicitly shows the benefit of being able to walk or cycle to work. Did people work at home because they thought it was more efficient or because they could not get to work? South Norfolk Council were preparing to run a skeleton service run by staff who could walk or cycle to work. This is further evidence of the lack of importance we currently attach to home location decisions relative to the activities we take part in.

The media had a very important role to play in the fuel shortages. Increases in the use of 24-hour news programmes and Internet sites as well as traditional news programmes showed that the UK wanted to be informed of how the crisis was going to affect them. Whilst the coverage certainly fulfilled this role it would seem that the reporting was at best selective in the impact evidence it presented, concentrating on the losers when clearly there were winners too. Very little follow up reporting was found in our searches, reflecting how the news bandwagon moved on to greater problems such as flooding, the Hatfield rail crash and Foot and Mouth. Whilst these were not issues that affected everyone at the same time in the same way that the fuel crisis did, they were much more sustained problems and perhaps more deserving of the title crisis.

When new drive-thru restaurants, multiplex cinemas and shopping centres open there are benefits and problems. The impacts are spread over the UK and over time but livelihoods are improved and at the same time threatened. Perhaps the fuel shortage has provided us with a unique opportunity to turn back the clock to a period of lower mobility and see how different things are. There were some winners and some losers during the shortage but at the end of it, UK Plc rolled on apparently unaffected by this blip. We conclude by answering the question posed by our title with drama, not crisis.

References

Ananova (2000a). Ananova Website. Available at: http://www.ananova.com/news/story/sm_60176.html[accessed 26/07/01].

Ananova (2000b). Ananova Website. Available at: http://www.ananova.com/news/story/sm_60995.insdat [accessed 26/07/01].

Ananova (2000c). Ananova Website. Available at: http://www.ananova.com/news/story/sm_61237.insdat [accessed 26/07/01].

Ananova (2000d). Ananova Website. Available at: http://www.ananova.com/news/story/sm_61424.html [accessed 26/07/01].

Ananova (2000e). Ananova Website. Available at: http://www.ananova.com/news/story/sm_61743.html [accessed 26/07/01].

Ananova (2000f). Ananova Website. Available at: http://www.ananova.com/sport/story/sm_62191.insdat [accessed 26/07/01].

Ananova (2000g). Ananova Website. Available at: http://www.ananova.com/news/story/sm_73572.html [accessed 26/07/01].

Ananova (2000h). Ananova Website. Available at: http://www.ananova.com/news/story/sm_76837.html [accessed 26/07/01].

Ananova (2000i). Ananova Website. Available at: http://www.ananova.com/news/story/sm_92669.html [accessed 26/07/01].

Ananova (2000j). Ananova Website. Available at: http://www.ananova.com/entertainment/story/sm_105236.html [accessed 26/07/01].

Ananova (2000k). Ananova Website. Available at: http://www.ananova.com/news/story/sm_123764.html [accessed 26/07/01].

Ananova (2000l). Ananova Website. Available at: http://www.ananova.com/news/story/sm_131527.html [accessed 26/07/01].

Atkinson, M. (2000). Spending spurs rate fears. *Guardian*, 20/10/00.

ATOC (2000). *Fuel Shortage Impacts on Performance*. ATOC Bulletin, October 2000. Available at: http://www.40bernardstreet.co.uk/publications/bulletin/ [accessed 20/06/01].

Baker, L. (2000). Waitrose feasts on fuel shortage panic buying. *Independent*, 23/09/00.

BARB (2001). All TV viewing figures from the Broadcaster's Audience Research Board Ltd website. Available at: http://www.barb.co.uk [accessed 26/07/01].

BBC News (2000). BBC News Online Website. Available at: http://news.bbc.co.uk/hi/english/world/europe/newsid_851000/851328.stm [accessed 26/07/01].

BBC Sport (2000). BBC Sport Online Website. Available at: http://news.bbc.co.uk/sport/low/english/newsid_926000/926035.stm [accessed 26/07/01].

BCC (2000). British Chambers of Commerce website. Available at: http://www.britishchambers.org.uk/newsandpolicy/downloads/FuelSubmission3.pdf [accessed 26/07/01].

BRC (2000a). British Retail Consortium website. Available at: http://www.brc.org.uk/news/rsm-oct.htm [accessed 26/07/01].

BRC (2000b). British Retail Consortium website. Available at: http://www.brc.org.uk/news/srsm-nov.htm [accessed 26/07/01].

Brown, K. (2000). Manufacturers prop up the dominoes. *Financial Times*, 22/09/00.

BT (2000). Btworking from home website. Available at: http://www.workingfromhome.co.uk/...etter/oldnewsletter/newsitemB.htm [accessed 26/07/01].

CAA/ACNielsen EDI (2000). CAA UK Advertising Admissions Monitor of all cinemas accepting advertising.

Caterer (2000). Caterer and Hotelkeeper magazine website. Available at: http://www.caterer.com/Archive/archivedetails.asp?ID=28353 [accessed 26/07/01].

Chatterjee, K. and Lyons, G. (2002). Travel Behaviour of Car Users During the UK Fuel Crisis and Insights into Car Dependence. In Lyons, G. and Chatterjee, K. (eds.) *Transport Lessons from the Fuel Tax Protests of 2000*, Ashgate, Aldershot.

Coach and Bus Week (2000). Operators Battle Against Fuel Shortages. September 21st, 3, Emap automotive.

Cozens, C. (2000). Tankloads of viewers. *Guardian*, 18/09/00.

Crooks, E. (2000). Consumer confidence in a slump after fuel crisis. *Financial Times*, 29/09/00.

Daily Express (2000). Fuel Crisis bit into store sales. *Daily Express*, 23/09/00.

Davison, P. (2001). Personal Communication. 25/06/01.

Denny, C. (2000). British manufacturing shrugs off hauliers' blockades. *Guardian*, 03/10/00.

DETR (2000). *New Directions in Speed Management – A Review of Policy*. Department of the Environment, Transport and the Regions, London, March.

DETR and EA pollution impact assessment study.

De Vlieger, I., De Keukeleere, D. and Kretzschmar, H.R. (2000). Environmental effects of driving behaviour and congestion related to passenger cars. *Atmospheric Environment* 34. 4649-4655.

Doran, J. (2000). Business will lose millions, says CBI. *The Times*, 15/09/00.

Dun and Bradstreet (2000). Dun and Bradstreet website. Available at: http://www.dnb.com/about/media/press_release/1,,0-222-1019-0-775.html [accessed 26/07/01].

EC (2001). *Towards a European Strategy for the Security of Energy Supply*. Green Paper, European Commission, Office for the Official Publications of the European Communities, Luxembourg.

Elliott, V. (2000). Panic buyers force stores to ration food. *The Times*, 14/09/00.

Evans, J. (2000). The price of petrol. *Sunday Telegraph*, 17/09/00.

Eves, D., Quick, J., Boulter, P. and Hickman, J. (2002). The Effect of the Fuel Crisis on Sections of the English Motorway Network. In Lyons, G. and Chatterjee, K. (eds.) *Transport Lessons from the Fuel Tax Protests of 2000*, Ashgate, Aldershot.

Felsted, A. (2000a). Securicor counts cost of fuel blockade. *Financial Times*, 28/09/00.

Felsted, A. (2000b). High street gloom descends on Oasis. *Financial Times*, 06/10/00.

Gascoigne, C. (2000). Fuel crisis hits weddings and holidays. *Financial Times*, 16/09/00.

GMTU (2000). *Some Effects of the September 2000 Vehicle Fuel Shortage on Traffic and Travel in Greater Manchester*. GMTU Report 686, Greater Manchester Transportation Unit, Manchester.

Grande, C. (2000). Online shopping supermarkets launch trials as delivery services gather pace. *Financial Times*, 16/09/00.

Griffin, R and Wilson, J. (2000). Special report: the petrol war. *Guardian*, 16/09/00.

Hathaway, P. (2001). The effect of the fuel 'protest' on road traffic. *Transport Trends*, TSO, London, 67-69.

HBAA (2000). Hotel Booking Agents Association website. Available at: http://www.hbaa.org.uk/Release_11.htm [accessed 26/07/01].

Hertfordshire County Council (2000). *An Investigation into the Effect of the Fuel Crisis of September 2000 had on Traffic Flows in Hertfordshire*. Hertfordshire County Council, Hertford.

House of Commons (2000). *Recent Events on the Railway.* First report of the Select Committee for Environment, Transport and Regional Affairs, December. Available at: http://www.publications.parliament.uk/pa/cm200001/cmselect/cmenvtra/17/1703.htm [accessed 26/07/01].

The Independent (2000a). Virtual sellers hit by real world oil crisis. *Independent*, 25/09/00.

The Independent (2000b). The Investment Column: 'Silentnight', *The Independent*, 28/09/00.

Investors Chronicle (2001). Smaller Company Annual Reports, 'Gowrings- motor dealer, fast food dealer', *Investors Chronicle*, 27/04/01.

IoD (2000). Institute of Directors website. Available at: http://216.31.194.212/is-bin/INTERSHOP.enfinity/eCS/Store/en/-/GBP/IODLeftNavigation-StartAdvice;%24sid%24ah9gO-kGYDvf0VVsUUgrqA956-N-BZ1W?Target=Main [accessed 26/07/01].

Kilgreen, L. (2000). Retailers blame fuel crisis. *Financial Times*, 27/09/00.

Kraft Burke, L. (2000). Accidentcare in profits slide. *The Scotsman*, 23/12/00.

Lambeth, J. (2000). Crisis hits Waitrose deliveries. *Daily Telegraph*, 14/09/00.

Lancashire County Council (2000). *Effects of Fuel Crisis Week in Lancashire.* Lancashire County Council, Preston.

Leeds City Council (2000). *An Assessment of the Impact of the Fuel Crisis on Travel Behaviour in Leeds.* Leeds City Council.

Loader, A., Willis, P. and Handley, C. (2001). *Investigation of the Effect of the September 2000 Fuel Shortage on Air Quality.* AEAT report AEAT/ENV/R/0477 report for the Environment Agency – Restricted, June.

London (2000a). This is Local London website. Available at: http://www.thisislocallondon.co.uk/archive/2000/09/14/enfieldnews3RM.html [accessed 26/07/01].

London (2000b). This is Local London website. Available at: http://www.thisislocallondon.co.uk/archive/2000/10/04/hounslownews2RM.html [accessed 26/07/01].

Marks and Spencer (2000). Marks and Spencer website. Available at: http://www2.marksandspencer.com/thecompany/news/pressrelease/fin2000-09-18-00.shtml [accessed 26/07/01].

Marsden, G. and Bell, M.C. (2001). Road Traffic Pollution Monitoring and Modelling Tools and the UK National Air Quality Strategy. *Local Environment*, 6(2) 181-198.

Mceachran, J. (2000). Record 70,000 crowd at Leuchars air show. *The Scotsman*, 18/09/00.

McRae, I.S., Green, J.M., Hickman, A.J., Hitchcock, G., Parker, T., and Ayland, N. (2000). *Traffic management during high pollution episodes: a review.* TRL Report 459, Berkshire.

Menary, S. (2000). Tesco beats fuel crisis as sales reach five per cent. *The Scotsman*, 29/11/00.

Menary, S. (2001). The gentleman who spends his money on The Officers Club. *The Scotsman*, 02/01/01.

Mid Sussex (2000). This is Mid Sussex website. Available at: http:www.thisismidsussex.co.uk/archive/2000/10/24/BUSINESS10ZM.html [accessed 26/07/01].

National Statistics (2000). National Statistics website. Available at: http://www.statistics.gov.uk/pdfdir/rsbrief1000.pdf [accessed 26/07/01].

National Trust (2000). Figures provided by Caroline Tse, Database Marketing, National Trust.

North East (2000). This is the North East website. Available at: http://www.theclarion.co.uk/the_north_east/archive/2000/11/07/A4gav.na.html [accessed 26/07/01].

Pain, D. (2000). Food firm shrugs off fuel crisis. *Independent*, 20/09/00.

Patten, S. (2000). JJB blames fuel crisis for drop in sales. *The Times*, 12/10/00.

Perkins, M. (2000). Fuel Crisis September 2000. Personal Communication. Serco Metrolink, Manchester.

Phelps, J. (2000). JJB scores poorer result after fuel crisis. *The Scotsman*, 12/10/00

Pook, S. (2000). Businesses crippled and job fears grow. *Daily Telegraph*, 14/09/00.

RAJAR/RSL (2000). Radio Joint Audience Research Limited website. Available at: http://www.rajar.co.uk [accessed 26/07/07].

Ryall, D.B. and Maryon, R.H. (1998). Validation of the UK Met. Office's NAME model against the ETEX dataset. *Atmospheric Environment*, 32(24), 4265-4276.

Schucker, M. (2001). Personal Communication. 25/07/01.

Scottish Tourist Board (2001). Figures provided by the Tourism Futures Department of the Scottish Tourist Board, for more information see website. Available at: http:www.scotexchange.net [accessed 26/07/01].

The Scotsman (2000a). Panic buying hitting food stocks. *The Scotsman*, 15/09/00.

The Scotsman (2000b). Business Staff: Food sales rocket during fuel crisis, *The Scotsman*, 23/09/00.

SMMT (2000). The Society of Motor Manufacturers and Traders Ltd website. Available at: http:www.smmt.co.uk/news/pressreleasedisplay.asp?articleid=228; http:www.smmt .co.uk/news/pressreleasedisplay.asp?articleid=231; http:www.smmt.co.uk/news/press releasedisplay.asp?articleid=242 [accessed 26/07/01].

Somerfield (2000). Somerfield plc website. Available at: http://www.somerfieldplc.co.uk/ index.asp?p=1&t=2&l=2&limit=off&release=94 [accessed 26/07/01].

South Norfolk Council (2000). *Cabinet Report – Fuel Crisis – feedback*. November, Long Stratton, South Norfolk.

Southern Vectis (2000). *Petrol Protests - Survey of Non-Bus Users*. Southern Vectis, Isle of Wight.

Taylor M., Lynam, D. and Baruya, A. (2000). *The effects of drivers' speed on the frequency of road accidents*. TRL Report 421, Crowthorne.

The Times (2000). Fuel crisis fails to dent high street spending. *The Times*, Business, 19/10/00.

TRANSIT (2000). Bus operators draw up contingency plans to cope with future fuel shortages as fuel crisis sees 20% surge in rail patronage. *TRANSIT*, 141, September.

Turner, D. (2000). Forecasters rethink impact of fuel crisis. *Financial Times*, 25/09/00.

Vnunet (2000). Vnunet website. Available at: http://www.vnunet.com/News/1110913 [accessed 26/07/07)

Whitaker, R. (2000). Day trips become luxuries as coach operators run out of fuel. *Independent on Sunday*, 17/09/00.

Whiteing, T., Coyle, M. and Bamford, C. (2002). Effective Fuel Management in Road Transport Fleets. In Lyons, G. and Chatterjee, K. (eds.) *Transport Lessons from the Fuel Tax Protests of 2000*, Ashgate, Aldershot.

Winpenny (2000). Charles Winpenny Cornish weather records website. Available at: http://www.winpenny.freeserve.co.uk/weather/weatherindex.htm [accessed 26/07/01].

Worcestershire (2000). This is Worcestershire website. Available at: http://thisis worcestershire.co.uk/archive/2000/10/04/wen_news_business02ZM.html [accessed 26/07/01].

Wright, M. (2000). Confidence endures the fuel crisis. Daily Telegraph, 23/10/00.

Appendix - Government List of Prioirty Users

The 29 priority categories of users identified by the Department for Trade and Industry in press release P/2000/621 were as follows:

1. Performance of functions of the police or the armed forces, or operation of the fire services.
2. Operation of health or social services.
3. Manufacture, processing, storage, packaging, distribution or sale of food including milk, milk products, animal feeding stuffs, pet foods or salt.
4. Movement or slaughter of livestock for the production of food or of animal feeding stuffs or for veterinary reasons.
5. Essential operations at nuclear sites.
6. Maintenance of water supplies or operation of sewage and drainage services.
7. Production, supply, distribution or transmission of solid or liquid fuel, or gas or electricity in the course of business, or provision of emergency services to consumers of such fuels.
8. Operation of buses, railways, tramways, docks or harbours.
9. Operation of any of the following transport services –

    - coach or bus services (excluding excursions or tours);
    - services provided by employers of labour for taking their employees to or from their work by means of vehicles adapted to carry eight or more passengers; and
    - airport bus services.

10. Operation of air traffic control, air navigation and civil airline services or air marine safety or navigational services.
11. Marine safety, navigational (including pilotage), shipping and ferry services.
12. Search and rescue services including coastguard and lifeboat services.
13. Production or distribution of medical or pharmaceutical supplies.
14. Operation of postal and telecommunications services (including broadcasting, newspaper production and distribution).
15. Transmission of money or of social security and other essential official payments or documents.
16. Movement or delivery of cash or cheques.
17. Operation of payment, clearing or settlement services.
18. Agriculture or horticulture for the production of food or animal feeding stuffs.

19. Operation of fishing vessels.
20. Manufacture or distribution of ice or dry ice for use in the food trades.
21. Essential central government and local government functions.
22. Administration of justice, including the operation of prison services and electronic tagging services.
23. Environmental and environmental health services and related emergencies.
24. Funeral services including cremation.
25. Refuse collection and disposal.
26. Industrial processes involving hazardous or radioactive substances, including transportation.
27. Veterinary services.
28. Operation (including heating) of special schools or colleges for the handicapped.
29. Foreign diplomatic missions for essential purposes.

The priority list came into force on Thursday 14th September 2000.

13 The Top of the Escalator?

GRAHAM PARKHURST

Reaching the Ceiling

Motoring taxation has been controversial almost since motoring began. When Asquith argued in 1907 that a tax on cars 'would be almost an ideal tax, because it is a tax on luxury which is apt to degenerate into a nuisance', he was greeted with the then dubious claim that cars were already a functional part of 'ordinary' people's lives (Plowden, 1971: 84). Thus was pioneered the crux of the debate which surrounded the fuel price protests of 2000, about the socio-economic necessity of motor vehicles versus the need for environmental protection from them.

Increases in tax rates, together with growth in vehicle ownership and road traffic, have been the key factors leading to real-terms increases in motoring tax revenues in recent years. The policy of fuel duty increases, which became known as the 'escalator', was not, though, the only cause of the fuel price rises in the service stations that led to the protests of 2000. Indeed, the United Kingdom (UK) Government sought to identify crude oil prices as the main factor and, arguably, the increase of crude oil prices from around $10 a barrel in September 1999 to peaks of $30-35 a year later was the most important immediate cause. Nonetheless, the public was made aware that, despite increases in product prices, taxation accounted for three-quarters of the costs of petrol and diesel to the consumer.

Environmental justifications for resisting reductions in fuel duty rates were also deployed. Some commentators argued, though, that raising revenue, rather than environmental protection, was the ultimate motivation. Government itself acknowledged that raising revenue had been a factor, by adopting a pragmatic defence that tax cuts would threaten important public expenditure programmes in the health and education sectors.

Nonetheless, in recognition of the limited political effect of these arguments, and accepting the reality of the vociferous protest that 'things had gone too far', government yielded in a limited way to political pressure. What could be termed a 'conciliation package' emerged between the 2000-2001 and 2001-2002 national budgets. The duties on the most widely used formulations of petrol and diesel were reduced by 3 pence per litre (around 6 per cent), and significant cuts in vehicle excise duty (VED) were made for targeted groups of vehicle owners.

The political need for tax concessions was perhaps an indicator that fuel prices had reached a threshold of acceptability. Such a ceiling would be critical not only for government revenue but also for government policy. The UK Government is committed to a reduction in 'greenhouse gas' emissions (equivalent to 12.5 per cent below 1990 levels by 2008-2012). Along with other EU member state governments, it has agreed to implement policies that will complement manufacturers' attempts to increase the efficiency of new cars sold. Furthermore, aside from the specifics of transport policy, the 1997-2001 Labour administration (re-elected in June 2001) was noted for its strong general preference for raising additional revenue from indirect taxation, rather than income taxation.

The emergence of the protests in 2000 and the uncertainties about crude oil prices raise a number of questions. Has the fiscal pressure on government motoring taxation policy now peaked? Are more protests likely in the foreseeable future? Which will be the future 'winner' and 'loser' groups? Will the motoring taxation system need to be adapted, not only in order to respond to protesters' concerns, but also to ensure stability of government revenues?

The analytical part of the present chapter begins by examining motoring taxation policies as they apply to car use. It is noted that, for each measure, a reduction in revenue (whether per kilometre or per car) can be expected in the future.

Next, the expectation that per unit revenues will reduce is placed in a wider context. Traffic growth can be expected to 'mature', i.e. slow down and possibly halt as: car ownership begins to approach saturation; traffic restraint policies are increasingly implemented; and road congestion increases. If, at the same time, vehicles become more efficient, then motoring tax revenues will peak and begin to decline. Calculations are made to determine the likely changes in tax revenue that would occur under five possible traffic growth scenarios. The effects of discount policies - which apply to certain VED liabilities and to duties on most fuels other than petrol and diesel - are taken into account.

The present circumstances of political constraints on increasing fuel duties may not, of course, be permanent. Budget decisions to increase duty rates could be a means of ensuring that revenues do continue to rise. The last two sections of the chapter consider the likely sources of controversy about motoring taxation in the future. It is noted that the range of efficiency exhibited by the car fleet is widening and it is argued that the cost of motoring is likely to increase, due to growing maintenance demands and the adoption of entirely new technologies that are potentially more expensive. As a result, some motorists are likely to become increasingly sensitive towards taxation policy, creating new potentials for conflict. Hence, increases in fuel duty may be problematic to implement.

It is concluded that, given both this context of contestability and the insecurity of fuel duty revenue, the government interest would be served by deriving a part of national motoring taxation from a new source, such as a system of tolls on trunk roads. Furthermore, if motoring is to become more expensive, and so more exclusive, then that places even greater emphasis on the need for satisfactory public transport alternatives to be provided.

Policies for Transforming Car Technology

The emissions control and traffic policy measures targeted at UK motorists can be classified into three groups, according to their motivations and mechanisms (Table 13.1).

Table 13.1 Classification of motoring policy regulatory measures

	Motivation		Mechanism		
Measure	Emissions control	Traffic management	Technological transformation	Reduction/ redistribution of vehicle-km	Group
1. VED reform 2. Fuel standards 3. Alternative fuel incentives	✓	×	✓	×	1
4. Road user charges 5. Parking charges	✓	✓	×	✓	2
6. Access restrictions 7. Fuel duty	✓	✓	✓	✓	3

As cars are both emitters of carbon dioxide (CO_2) and noxious pollutants, all seven measures have a link with emissions policy. However, the extent of this relevance varies. For example, access restrictions in town centres are most likely to reduce air pollution concentrations, but if they encourage modal transfers to public transport then they may also reduce CO_2 emissions. Similarly, a new graduated VED measure has been designed mainly to reflect the level of CO_2 emissions from a particular car, but also incorporates a secondary adjustment according to fuel type, intended to reflect the level of noxious emissions.

Four measures in Table 13.1 have a direct influence on traffic. It is notable that the policy for high petrol and diesel duties is the only one that has been implemented to a significant extent.

National legislation for new motoring charges - road-user tolls and levies on workplace parking spaces - has been enacted in Part III of the Transport Act 2000 (HM Government, 2000). These charges will only be introduced on a local basis, at the instigation of the relevant local authorities. To some extent they could perform the same function as the traffic restraint element of fuel duty policy. However, given the significance of fuel duty revenue to government finances, it is important to note that the net proceeds will be hypothecated to local authorities. Hence, to the extent that these new charges reduce car use, they would be contrary to Treasury interests. New legislation would be necessary to provide for charging schemes that could benefit the Treasury. But, in principle, national road tolls could either contribute to Treasury income or be hypothecated for specific purposes, or be apportioned to both of these ends.

Traffic restriction regulations are growing more common. The implementation of local air quality strategies will be necessary in some relatively polluted locations in order to meet national objectives for 2005. These strategies may involve traffic restrictions that discriminate between vehicles according to their emissions levels (DETR et al., 2000).

In contrast, the Group 1 measures of Table 13.1, which only target emissions, have been widely implemented. Vehicle and fuel type regulations have been negotiated at the European level. On-going implementation of air quality measures under the Auto-Oil programme began in 1993. In 1998, a voluntary agreement was reached with manufacturers to reduce CO_2 emissions by 25 per cent by 2008. The EU governments also agreed to introduce higher fuel duties to encourage consumers to buy these more efficient cars, but the UK Government has pursued that policy particularly firmly. Indeed, with hindsight it can be argued that the policy of implementing 'automatic' 5-6 per cent above-inflation increases for much of the 1990s was too radical. Rises in crude oil prices in fact led to the 'escalator' policy being suspended in 1999. The protests that nonetheless ensued constitute evidence that the main policy deployed to restrain traffic growth was in fact implemented to the limit of political acceptability.

Incentive measures have now replaced 'deterrence'. In 1999, a twin-band VED regime replaced the single-rate charge for UK cars, with the intention of transferring a proportion of car travel costs from ownership to use. Initially only cars with engine capacities of 1,100cc or less were eligible for the £55 per annum discount, but, as noted in the introduction, by the time of the 2001-2002 budget the discounted band had been extended to cars with engines of up to 1,549cc. These cars constitute around half of the total fleet (DETR, 2000d: Table 3.3). Additionally, from 2001, new cars became subject to the graduated VED scheme noted in Table 13.1. The regime has four bands, with a range of £70 between the highest-taxed and lowest-taxed cars.

The UK Government also offers conversion grants of up to 75 per cent of the additional cost of equipping new vehicles to operate on fuels other than petrol or diesel, and provides fuel duty discounts for road fuel gases (RFGs). Following the 2001-2 budget, the combined fuel duty and Value Added Tax (VAT) paid by private users of Liquefied Petroleum Gas (LPG) was reduced to 25 per cent of that paid for comparable petrol cars (Parkhurst, 2001: Table 5). At the time of writing, Compressed Natural Gas (CNG) has a very small market share as a car fuel in the UK, but the implied discount is at least as great. Electricity has an even more favourable position, subject only to 5 per cent VAT.

Implications of Emissions and Traffic Policies for Tax Revenues

The main reason why government was able to rely on rising revenues from motoring throughout most of the 20th Century was increasing traffic. A secondary factor is that the specific fuel consumption of UK cars has reduced at a slower rate than traffic has increased. Dargay and Gately (1997: Table 4) have estimated that demand for car fuel in the UK will increase by 23.5 per cent 1995-2015. Their view is based on assumptions, *inter alia*, that car traffic increases by 38.5 per cent in the same period and that average car fuel consumption reduces by 11 per cent. The actual development of these two trends will clearly be critical for revenue.

In reviewing progress under the Road Traffic Reduction (National Targets) Act 1998 (HM Government, 1998) and during development of the *Ten-Year Plan for Transport* (TYP), the UK Government Department of the Environment, Transport, and the Regions (DETR) considered restraint scenarios that would limit traffic growth to between 8 per cent and 17 per cent during the decade 2000-2010 (DETR, 2000a, b, c). In the absence of new restraint measures, traffic growth is forecast to be 22 per cent. DETR scenarios include the assumption that crude oil prices will decline over the decade from $28 to around $16 a barrel. Such a view is vulnerable, though, to unpredictable international events. It also assumes that oil production will keep pace with growing world demand through the next decade. Some geologists argue that the point at which easily extractable oil reserves are no longer available to exploit is being reached, and that prices will inevitably rise afterwards. (Campbell and Laherrier, 1998). Magoon (2000) reports that geologists' estimates about the date when demand for oil will exceed world production capacity - the 'big rollover' - range from 2003 to 2020. Magoon suggests that the approach of the 'big rollover' is likely to keep crude oil prices above \$25 a barrel, which also happens to be the central value of the OPEC target price range; reductions in oil prices below \$25 triggered three OPEC production cuts in the first seven months of 2001. Higher average oil prices over the decade would tend to result in lower traffic growth.

Improved fleet efficiency as a result of successful fulfilment of the agreement between the EU and car manufacturers would result in a 26 per cent reduction in fuel consumption by new UK cars by 2008. If the part of the agreement subject to review in 2003 is confirmed then a further target of 36 per cent reduction by 2012 will be set (Parkhurst, 2001: Appendix 2).

Doubts have been expressed about the ability of car manufacturers to achieve the efficiency target unless the average specification in terms of weight, size, and engine power is reduced (Kågeson, 2000: 34-5). There are also concerns about whether the official test data for particular car models in practice translate into similar real-world reductions (Van den Brink and Van Wee, 2001: 86-91). However, differences between existing models of similar size and power output (TRI/ECU, 2000: 94), and similar size and engine capacity (Sakaguchi, 2000: 898), can be greater than 20 per cent. Sakaguchi (2000: 903) found that the technological measures proposed as part of a very similar efficiency improvement strategy for Japan are potentially adequate. Actual fulfilment will depend on the rate of diffusion of the new technologies, which itself is a function of speed of technological development and consumer acceptance. Notably, though, even under the highest traffic-growth scenario (i.e. with a low level of adoption of new technologies), demand for petrol by Japanese car users is expected to peak in 2010. If, instead, a strategy for faster diffusion is successful, fuel demand will peak in 2005 and begin to decline, returning to 1999 levels in 2013.

Similarly, effective policies will be necessary if the European targets are to become reality. The nature of the revised UK VED regime shows commitment to this end. It is noted, though, that nearly all vehicle-fuel combinations result in a lower duty being paid.

The extent to which British motorists will respond to the opportunity to avoid high fuel taxes by adopting alternative fuels is uncertain. The most practical alternative at the time of writing was LPG, which is usable in adapted cars that were originally designed to run on petrol. LPG is still only sold in a minority of UK service stations, but availability is growing. Refuelling and on-vehicle storage of CNG is somewhat less straightforward, but is particularly suited to heavy vehicles designed to operate using diesel. The existence of a national pipeline network represents an important asset. The same is true for electricity. There have been important developments in electric vehicle technology in recent years, and a number of small citycars now represent a practical alternative for the minority of motorists who know they will only ever want to use their vehicles for fairly local journeys. Further advances in battery recharge speed and storage performance would, however, strongly increase the competitive position of electricity. In the meantime, more radical change is likely to depend upon the exploitation of hybrid (internal combustion engine combined with an electric motor) and fuel cell technologies.

The potential for change is revealed by evidence that, by 2000, LPG alone had already achieved market shares of 4 per cent in Italy, 5 per cent in Australia, and 6 per cent in the Netherlands (Parkhurst, 2001: Table 4). Similarly, the efficacy of tax discounts is revealed by the growth of diesel car sales in the EU. An average 35 per cent duty discount for diesel compared with petrol in the EU states (except the UK) has resulted in a doubling in the market share of diesel cars in the EU to 25 per cent in the last decade (Kågeson, 2000: 10, 17). These statistics indicate that financial motivations can strongly influence consumer choice, provided that the alternatives are realistic.

The development of alternative technologies is unlikely to be a 'zero-sum' game in which advances in respect of one approach eliminate the development of the others. At the same time, they will not be entirely mutually exclusive; the markets for the different technologies are likely to be overlapping. Each has valuable attributes, which suggests that particular technologies will be influential at different points in the move away from dependence on oil. At present, greater availability of RFGs in the UK, particularly in the case of LPG, is making the operation of bi-fuel vehicles a realistic consumer choice. Further impetus to adopt RFGs will follow once fuel supply networks are sufficiently developed so as to render the higher vehicle construction cost of bi-fuel capability unnecessary.

In every case, though, as the market share increases, a combination of greater efficiency and government policies offering large duty discounts will begin to result in non-trivial reductions in revenues. Hence, it can be hypothesised that, taken together, the policies for reducing the environmental effects of car use will have unavoidable policy implications for government, which will require changes in transport policy or taxation policy, or both.

The following section tests that case by estimating the likely extent of change. For the sake of analysis, the case of RFGs is selected to represent the growth in alternative technologies over the next decade. In practice, the same dilemma will be created if another technology is significant within the market for alternatives, except that the revenue reduction consequences are likely to be greater.

Estimation of the Magnitude of Revenue Changes

The estimations presented refer to the levels of fuel and vehicle excise duties that will be paid by car owners and users over the next decade. Income tax measures (company car and fuel benefit taxes) are not considered. Efficiency and duty rate changes affecting road freight transport will also have implications - Chapter 10 (Whiteing et al., 2002) considers the potential for efficiency responses within that sub-sector - but the present analysis is concerned solely with car traffic, the largest contributor of motoring taxes (DETR, 2000d: Table 1.20). Values and sources for 2000 base-year data are indicated in Table 13.2. Traffic and revenue data refer to Great Britain rather than the UK (i.e. Northern Ireland data are excluded).

Five traffic growth scenarios, summarised in Table 13.3, are selected from those considered by the DETR (2000a, b, c) for 2000-2010. The DETR scenarios allow for the elasticity of demand with respect to motoring costs, particularly fuel product prices and vehicle efficiency.

For interim years, incremental increases are assumed to be numerically equivalent (i.e. growth at a declining rate). This approach reflects three expected trends: the gradual introduction of new restraint measures over the decade; progress towards saturation car ownership amongst the wealthiest households; and the DETR assumption that per annum economic growth will be slightly lower in the second half of the decade. The analysis assumes that the car manufacturers achieve the agreed efficiency target for 2008 and make further progress towards the 2012 unconfirmed target. The rates and bands of the VED regime, the levels of fuel duties, and the relative discounts offered for RFGs are also assumed to remain unchanged, other than for inflation-linked increases. RFG vehicles, which in practice are likely to be LPG-powered in the majority, are assumed gradually to increase to account for 10 per cent of cars in 2010.

Based on data presented in the National Road Traffic Forecasts for Great Britain, the approximate rate of increase in car ownership in each scenario is assumed to be two-thirds of the rate of increase in traffic (DETR, 1997: Tables 1, 8).

Table 13.2 Assumptions for analysis baseline year (2000)

Assumptions	Value	Comment
Car and taxi traffic (billion vehicle-km)	3.8	Source: DETR, 2001: Table 2
Car fleet (millions)	23.24	Source: DETR, 2000d: Table 3.1 +2% growth 1999-2000
Car fleet average efficiency (gCO_2 emitted/km)	187.9	CO_2 emissions in 2000 of 188 g/km (Parkhurst, 2001: Appendix 1); largely unchanged for previous 15 years (DETR, 2000d: Tables 2.4, 2.5)
Income from fuel duty (£ billion)	14.97	Source: DETR, 2000d: Table 1.20 1998-9 income from private and light goods vehicles factored to discount the proportion travelled by light goods vehicles (14.5% based on DETR, 2000d: Table 4.9), inflated by 3.4% increase following 1999 budget.
VAT income from fuel duty (£ billion)	2.62	@ standard rate of 17.5%
VAT on fuel products (£ billion)	1	In practice liable to fluctuation; estimate based on pump price of £0.78/l (giving product price of £0.16/l)
VED paid by car owners (£ billion)	3.51	Average charge of £150 per car given dual-band system for 2000-1 financial year (10% of cars rated at £100 rather than £155)
RFG discount	69 %	Source: Parkhurst, 2001: Table 5

Table 13.3 Traffic growth scenarios considered in analysis

Scenarios	Growth 2000-2010 (%)
'Do nothing extra'	22
Basic measures of ten-year plan (DETR, 2000b)	17
Basic plan with constant motoring costs (DETR, 2000b)	13
Basic plan with new motoring charges (DETR, 2000b)	12
High charges + land-use measures (DETR, 2000a)	8

Other than increases in fleet size, resulting in the average age becoming somewhat younger, and the aforementioned increase in efficiency, no other assumptions about the cars are made. Increases in car ownership will have a positive effect on the rate of reduction of the average fleet consumption. A cohort analysis was undertaken to estimate this effect.

Most UK cars are scrapped when between 14 and 17 years of age (Commission for Integrated Transport, 2000: 9), and the present analysis makes the assumption that they are scrapped once (and only when) they reach 15 years of age. Each new cohort is assumed to have the average efficiency in the particular notional purchase year that is necessary for overall steady-rate improvement towards the dated targets. Each new cohort is also larger than the previous one, according to the level of growth assumed in each scenario. Hence, in the case of unrestrained car ownership growth (i.e. 15 per cent growth 2000-2010), the programmed efficiency improvement for new cars would result in a 9 per cent increase in whole fleet efficiency by 2008, compared with an 8 per cent increase under zero car ownership growth conditions (Parkhurst, 2001: Appendix 2).

Beyond 2008, as cars purchased prior to 2000 become a minority, and even without any further improvement for new cars, the fleet-wide average will reduce rapidly if the targets are achieved. By 2015 the fuel consumption reduction would compare with the 23 per cent reduction shown by the UK car fleet over the period 1980-1994; a change motivated by the 1970s oil crises (Orasch and Wirl, 1997: Figure 5).

Figure 13.1, charts the revenue changes for the three motoring taxes, considered individually and in combination, for the different levels of traffic growth.

Under the scenario of 17 per cent traffic growth (basic TYP), fuel tax revenues taken alone are in fact estimated to be higher in 2010 than in 2000. However, scenarios with greater traffic restraint would result in a reduction in fuel tax revenue.

By 2010, average per-kilometre CO_2 emissions of cars sold from 2001 will have reduced. The average car will move from the highest of the four emissions-related VED bands to the second lowest, resulting in a 20 per cent decline in the average VED paid. (It is assumed that current discounts will continue for the smaller-engine cars bought prior to 2001). The revenue implication is a reduction under all levels of car ownership growth.

The revenue from the use of RFGs is derived per kilometre, rather than per car. Hence, the reduction due to the discount is greatest where traffic growth is greatest. Under the 'do nothing extra' 22 per cent growth scenario, the discounts for RFG use would be equivalent to the increases in revenue from petrol and diesel duty that would be experienced due to extra traffic.

Taken together, the changes result in a net reduction in the range £1.25-2.75 billion at 2000 prices if traffic restraint is applied.

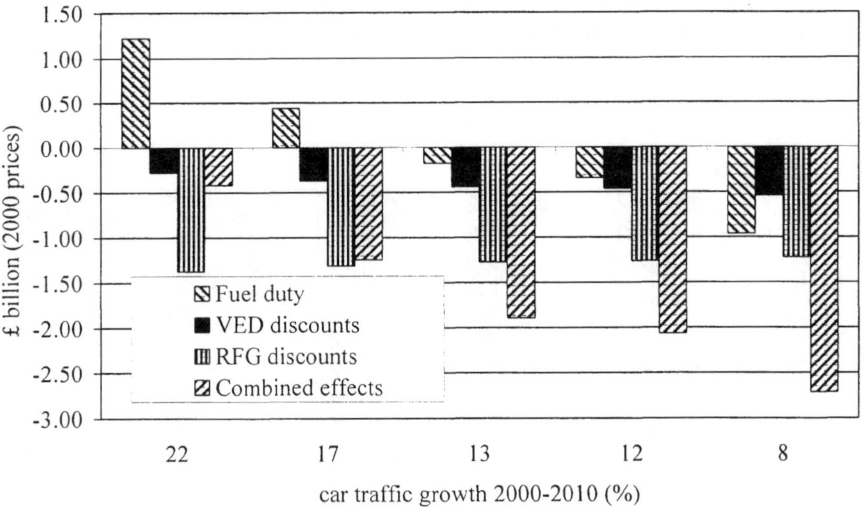

Figure 13.1 Change in motoring tax revenues from car use in 2010 under different traffic growth scenarios

Figure 13.2 extends the estimations to 2012. For the purposes of analysis, a prediction by former Secretary of State for the Environment, Transport, and the Regions John Prescott is tested, namely that:

> If the measures in the white paper are applied very intensively, we would be close to halting growth in national traffic levels in ten years' time, while delivering reductions in traffic to below 1996 levels in our major cities. (DETR, 2000a: 4)

In addition to zero traffic growth beyond 2010, it is assumed that the unconfirmed objective for average new car CO_2 emissions of 120g/km in 2012 is adopted and achieved. As before, VED and fuel duty rates are assumed to remain constant. No assumption about further increases in the use of alternative fuels 2010-2012 is incorporated.

In a context of no further traffic growth after 2010 and further improvement in car efficiency, the natural consequence is a greater reduction in fuel duty receipts than was estimated for 2010. If stabilisation occurs after 8 per cent traffic growth in the period 2000-2010, reductions in fuel duty due to greater efficiency would be the most significant of the three changes in receipts 2000-2012. The contributory explanation for that outcome is that the revenue 'cost' of RFG discounts falls slightly 2010-

2012; improved efficiency will also affect RFG vehicles, so the discount applies to a smaller quantity of fuel consumed.

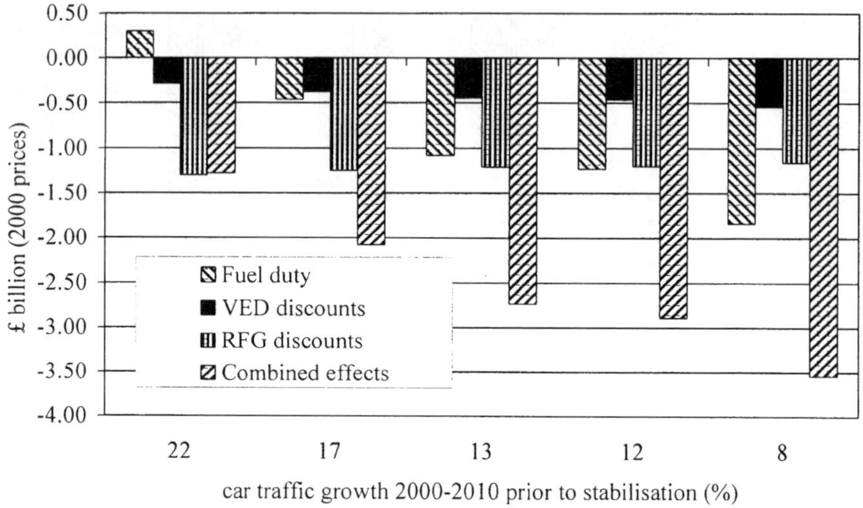

Figure 13.2 Changes in motoring tax revenues from car use in 2012 given stabilisation of traffic levels in 2010

VED revenue shows no change with respect to 2010. The tax loss in 2012 is not smaller than 2010 because the absence of traffic growth means that the car fleet size is assumed not to change. However the loss is not greater, either, because the incremental efficiency improvement 2010-2012 will not be sufficient to change the allocation of the vehicle fleet with respect to the CO_2-related duty bands in a significant way.

Combined, the estimated revenue implication, given implementation of the TYP with sufficient success to stabilise traffic levels in 2010, is for a reduction in the range £2-3.5 billion. Figure 13.3 indicates the expected evolution in net total revenue from the three sources.

Assuming reasonably smooth implementation of the basic TYP, revenue would peak by 2003 and decline to 2000 levels by about 2005. Notably, even in the absence of the TYP, 22 per cent traffic growth is insufficient to maintain revenue throughout the decade.

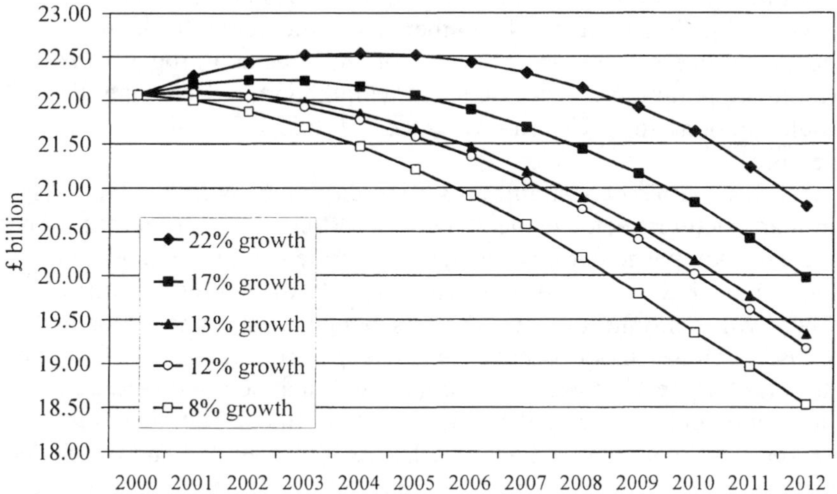

Figure 13.3 Evolution of motoring tax revenues under different traffic growth scenarios

Scope to Increase Duty Rates

Naturally, government would have the option to respond to eventual revenue reductions by raising duty rates. In each of the three cases, though, there will be constraints on that course of action.

Alongside the tax reduction measures of 2000-2001, government committed not to make real-terms petrol and diesel duty increases until the 2002-2003 budget at the earliest (HM Treasury, 2001). Hence, a political 'guarantee' exists in the short-term. There are also particular practical constraints on raising duty rates in the case of diesel. If UK prices were simply converted into Euros, then diesel has been up to 50 per cent cheaper in other EU states in the recent past. The real-terms cost to consumers buying in Euros in those other countries was not in fact as low; the strength of Sterling contributed to the creation of a 'cross-border' refuelling benefit, which could be exploited by travellers from the UK. Road hauliers were reported in June 2001 to be taking advantage of special 'fuel cruise' ferry trips to France. After buying a return ticket for £110, lorry drivers could save up to £190 by replenishing their 1000-litre fuel tanks. Such a possibility would not be open to car users, due to the much more limited capacity of their tanks, but diesel car drivers are directly affected by duty decisions taken primarily with the road haulage industry in mind.

At the same time, there are also policy reasons not to reduce diesel duty rates. Uniquely amongst EU member states, the duty rates levied on petrol and diesel are similar, in reflection of the greater nitrogen dioxide and particulate pollution levels associated with the latter. Reducing diesel duty would amount to a U-turn in respect of this air quality improvement measure.

By 2010, a new constraint on setting duty rates will develop. Whether car manufacturers achieve the 25 per cent efficiency improvement or not, it is clear that particular high-efficiency models will be made available. Some motorists will be driving cars capable of 3 litres/100km or better. Others will continue to drive vehicles bought in the late 1990s and early 2000s. These models have an average consumption of around 8 litres/100km, nearly three times higher, and include middle-range two-litre capacity models requiring 9-10 litres/100km. In contrast, there are plans to market diesel cars that can achieve 1 litre/100km by the end of the decade (Commission of the European Communities, 2000). There is also the prospect of fuel cell vehicles ultimately being powered by hydrogen derived from renewable sources.

Changes in fuel product price or duty rates already affect operators of relatively efficient vehicles less in absolute terms than those with relatively inefficient vehicles. The greater average efficiency of future new cars will increase the range of that variability, creating particular 'loser' groups. Raising petrol and diesel duty rates to maintain revenue in the context of declining total fuel demand would mainly result in higher tax costs for those with the least efficient cars; arguably effective from the perspective of environmental policy, but politically contentious in terms of social justice.

In the case of graduated VED, the rates for the bands could be increased above the rate of inflation. However, the view that the relative cost of car use with respect to car ownership should be increased has enjoyed a wide consensus. Moreover, in practical terms, changes in VED rates will have a relatively small influence on motoring tax revenue. The controversy that would be caused by a high profile 'U-turn' might not be worth the gain.

RFG discount policy, on the other hand, is estimated to be significant for revenue. Climate change politics in fact support a case for higher RFG fuel duty. The European Commission has proposed a road fuel taxation strategy that relates the CO_2 emissions per unit of equivalent consumption of a fuel directly with the rate at which it is taxed. RFGs do produce less CO_2 per kilometre than petrol, but the reduction is of the order of 25 per cent, whilst the duty discount is now 75 per cent. However, the UK Government has recently reiterated its resistance to energy tax harmonisation measures. In the 2001-2002 budget it committed not to increase duty on RFGs in real terms before 2004 at the earliest, in order to give confidence to the fledgling market (HM Treasury, 2001: 112).

Air quality policy objectives also urge the retention of high discount levels, as the proportions of noxious pollutants avoided by using RFGs instead of petrol and diesel are of a similar order to the discount. Moreover, the fuel product costs of RFGs are actually higher than petrol or diesel under present market conditions, once relative consumption is taken into account. Table 13.4 indicates that, for eleven LPG models produced by GM, Fiat, and Volvo, fuel costs per kilometre are one-third less than for equivalent petrol models. However, the product cost element for LPG is actually greater than for petrol. A similar situation applies in the case of CNG.

Individual consumers would generally not voluntarily accept these higher costs in order to create the public benefit of lower pollution. As the majority of cars that are intended to run primarily on RFGs also have the possibility to use petrol, motorists would be able to switch between the two, depending on a combination of convenience and price factors. The disincentives for using a non-standard fuel, for example more restricted availability than petrol and specialised insurance in some cases, suggest that a significant tax incentive will be necessary for many years to come if RFGs are to be attractive.

Table 13.4 Fuel product and tax costs in 2000 for 11 pairs of equivalent petrol and LPG models

	Petrol (@£0.84/l)	LPG(@£0.40/l)
Consumption (l/100km)	8.50	11.58
Cost to consumer/100km (£)	7.14	4.63
VAT/100km (£)	1.25	0.81
Duty/100km 2000-1 rates (£)	4.15	0.87
Product cost/100km (£)	1.74	2.95

In summary, the constraints surrounding motoring taxation policy are likely to increase rather than reduce through the decade 2000-2010. The proliferation of alternative technologies will make taxation decisions more complex, as motorists will have some possibilities to change fuel, according to which is the most tax-efficient option.

In the round, the ability of government to increase motoring taxation probably depends on the ability of citizens to pay the higher rates. The final section of this chapter argues, though, that developments in the automotive industry, which run parallel to changes in vehicle efficiency and taxation, will reduce the ability of particular groups of motorists to accommodate increases in motoring costs, including motoring taxes.

Technological Change and the Affordability of Motoring

Hutton (2000) has argued that fuel duty is a fundamentally regressive tax, as it is paid at the same rate by all motorists, but represents a larger share of poorer motorists' incomes. In practice, poorer motorists have the option of minimising their taxation burdens by modifying the amount they travel by car and by choosing the most efficient cars they can afford. *In extremis*, if motoring costs grow with respect to incomes, a few of the poorest motorists give up car ownership altogether. However, current trends suggest that a period is beginning in which the affordability of motoring for the poorest will reduce sharply. The implication is that a group of 'motorists at the margin' will emerge as a new source of constraint on motoring taxation policy.

Bolton and Twigger (2000: 11-12) indicate that the motoring elements of the retail prices index increased by 5.6 per cent between 1987 and 2000. As disposable household income per head also grew, i.e. by an average of 2.3 per cent per annum since 1979 (Institute for Fiscal Studies, 2001: 1), the average household is spending a similar proportion of its budget on motoring as it did in 1970 (Office for National Statistics, 2001).

However, although the average household in each income quintile has benefited from income growth, the poorest quintile has gained least, with the result that social inequality has grown since 1979. Furthermore, inequality has risen in the period 1997-2001, despite relatively low unemployment (Institute for Fiscal Studies, 2001: 4-5). One reason for the increase is that direct and indirect taxation taken together have grown in recent years for all households except those in the highest quintile of income. Taxes account for a larger share of lowest quintile incomes than they do of highest quintile incomes: 41.4 per cent compared with 36.5 per cent (Ward, 2001). Most (70 per cent) of households in the lowest quintile do not own cars, but motoring taxes are one source of indirect taxation for those that do.

Given this recent history, there is little reason to suppose that technological change intended to reduce the negative environmental consequences of road transport will itself be a force for greater social inclusion. Indeed, there are reasons to believe it is likely to be a force for greater inequality, which may manifest itself as political resistance.

National transport planning (DETR, 2000b, c) assumes that the UK economy will grow at annual rates of over 2 per cent per annum in the next decade, whilst the greater fuel efficiency of new cars will reduce the average fuel costs for motorists overall. However, historic trends suggest that the greater wealth implied will not be evenly shared. Those households that experience most growth in income will be more able to afford the more efficient technologies.

The possibility of the more efficient technologies being 'out of reach' is a specific example of the more general circumstance that motorists on low incomes have less choice about the cars they drive than richer ones. The pool of available vehicles is typically comprised by cars bought a decade earlier, under different market conditions. In recent years, more than half of cars first registered were purchased by companies (DETR, 1998), which apply different decision criteria to private purchasers, generally buying larger cars with bigger engines. In the second-hand market, the supply of these relatively large cars contrasts with relatively high demand from households, that perhaps already have at least one car, for small used cars to perform light duties. Hence, larger vehicles are oversupplied into the second-hand market, with the result that smaller cars with relatively low original purchase prices tend to depreciate less than larger cars with higher fuel and maintenance costs. In consequence, lower income buyers have less scope to 'downsize' to relatively small, efficient, cars than purchasers with more money to spend.

Furthermore, whilst overall motoring costs have probably increased more slowly in statistical terms than lowest quintile household incomes, not all of these costs apply equally to all households. In practice, the elements that affect 'budget' motorists most have increased much more sharply than the average. Insurance costs, which showed the greatest proportional increase 1987-2000, and fuel costs, a close second, have both risen by an average of more than 25 per cent in real terms. Maintenance costs have risen by around 20 per cent. At the same time, the largest decline, of around 20 per cent, occurred in respect of vehicle purchase costs. This, however, offers little benefit to lower-income motorists who tend to buy cars with minimal residual value. By the time proportional decreases in new car prices filter down through the used-car market, there is likely to be little benefit in absolute terms. In any case, most second-hand car prices, and particularly those of small and medium models, increased in the year to July 2001. There were also indications in the same month that the fall in new car prices had begun to reverse (Daily Telegraph, 2001a, b).

Moving into the future, technological development suggests upward pressure on new car prices. Lovins and Lovins (1996: 80-2) have argued that a radical revolution in the nature of car design and construction could result in cars which are much more efficient at no extra cost to the consumer. However, over the next decade, technological change is likely to be more incremental.

Aluminium is being increasingly used in cars to reduce weight, but is more expensive than steel. A lightweight version of a 'supermini' model launched in 1999 incorporated extensive use of aluminium, along with other weight and energy-saving technologies. Although some equipment differences limited direct comparability, it was tested as 50 per cent more efficient than the standard model from which it was developed. It had, though, a £1,000 higher list price (English, 1998). Hybrid-drive cars are already available and have two motors not one, as well as sophisticated batteries. Experience with the first production models shows that, even after large development subsidies from manufacturers (Bulgin, 1998; Hudson, 2001) and government incentives worth 40-75 per cent of the additional cost compared with conventional cars, the consumer can expect to pay a 'novelty' premium. These premiums reflect both the intrinsic nature of the technology and the initial low-volume production runs, and are likely to characterise the new solutions for some time to come.

A similar situation is observed with respect to maintenance costs. Catalytic converters were mandatory on all UK cars registered from August 1992. Seven years later, a national newspaper carried out an investigation of readers' experiences of the reliability of the new technology. The systems, although growing more robust as a result of gradual refinement through design evolution, were found to have proven very sensitive to chemical and physical damage and to have been associated with a high rate of failure. The cost of rectifying failures in order for the vehicle to pass the statutory annual emissions test (usually requiring replacement of the converter unit) was found to range between £200 and £2,000, although the cost of replacement units for small to medium models tended to lie towards the bottom of that price range (Daily Telegraph, 1999).

Other new technologies can be expected to create similar problems. The batteries implemented in hybrid and electric-only drive solutions will be expensive to replace and are of uncertain longevity. The manufacturers claim a life expectancy of ten years or the 'lifetime' of the car, but warranties are for four or five years (Hudson, 2000; 2001). In any case, in practice, motorists often run cars long beyond the 'expiry date' that suits the interests of manufacturers keen to maximise sales.

Perhaps of even greater concern than the early obsolescence of motive power components is the argument that the 'active' safety equipment now included in most new cars will also deteriorate, at an uncertain rate. Airbags were first available in UK cars in 1993. Manufacturers recommend that optimal performance can be expected for around ten years of life, so the earliest examples are now approaching their maximum usable age. If they are not systematically checked and replaced as necessary, they may either fail to function effectively, or even deploy during normal driving. In either case, the validity of insurance policies may be affected. The cost of replacing individual bags can range from £100-600, but given that most new cars now contain a number of airbags, the full replacement costs would regularly be expected to exceed £1,000 (Foxall, 2001). In practice, it is more reasonable to expect that cars will either be scrapped if the airbags need replacing or operated without replacements.

Cars with catalytic converters and airbags are currently entering the lowest price tier of the used-car market. As well as being additional components that are expensive to replace, they also tend to represent additional 'tamper proof seals'. Hence, the new equipment limits the extent to which low-income motorists can extend both the working life of cars and their affordability by practising amateur maintenance. The elimination of these older cars through built-in obsolescence may seem to promote the adoption of the best technology available for the environment, but these cars tend to be driven far less than newer ones, many of which form part of fleets operated by commercial organisations. Hickman (reproduced as Figure 3 in Zachariadis et al., 2001) estimates that UK cars are driven an average of 23,000 km in the first year of life. By the time they are ten years old, this distance has declined by nearly half, to 13,000 km, and to 7,500 km per year a decade later. Under certain circumstances, then, the driver using a relatively inefficient car less frequently may produce fewer climate change emissions per year than a driver with a relatively efficient car who travels much further.

At the same time as ownership costs are increasing, the introduction of additional restraint charges, for road use or parking or both, can be expected. As noted in Table 13.1, not all efficiency and pollution control measures reduce traffic. Indeed, some may actually encourage growth. Arguably, the case for new fiscal interventions to reduce traffic congestion is increased by greater efficiency, and especially if combined with falling tax revenues.

Taken together, these changes suggest that those who can only afford motoring by a small margin will be faced with more expensive cars and higher maintenance costs. Refuelling can be expected to cost as much, or more, if poorer motorists find it hard to keep pace with technological change. In addition to motoring becoming more expensive, the relative purchasing power of poorer motorists may fall. As fiscal policy measures will generally be aimed at the average income, poorer motorists may experience above average pressure.

For those unable or no longer able to afford car ownership, the option of public transport, of course, remains. Indeed, it is possible that, against a background of higher motoring costs, there will be greater concern for those without access to a car and public transport will be made more available and more affordable. If so, then car ownership will become less essential. That may prove to be the case, but recent history does not give much support to that hypothesis. Private sector operators and public sector policy makers have made considerable attempts to attract motorists to public transport in recent years. However, as most motorists are assumed to place convenience above cost, these attempts have usually been based on service quality enhancements, rather than permanent price reductions. Notably, there has been a 20 per cent real-terms increase in public transport fares between 1987 and 2000 (Bolton and Twigger, 2000: 12), a rate four times greater than the rise in overall motoring costs. The future agenda of better services and reconstruction of the railways suggests a continuation of this trend.

If government is prepared to redress growing inequality in the transport sector, then the implication seems to be that greater subsidies, probably directly applied to a wide range of fares, will be necessary.

Conclusions

The chapter opened by raising questions about future constraints on motoring taxation policy and the potential for further socio-economic strife. Farmers and the road haulage industry led the protests of 2000. Without denying that they expressed genuine grievances, it can be observed that by using their possibilities to negotiate tax relief with government, pass costs on to their customers, and invest in new technological solutions, these commercial interests will ultimately achieve realignment between transport costs and profits. Nonetheless, analysis of the policy mechanisms that contributed to the emergence of the fuel tax protest suggests that the issues will continue to be controversial, although in different ways, in the foreseeable future.

There will be a wider choice of vehicle efficiency and performance, and a greater choice of fuels. There will be the possibility for some individuals to reduce motoring costs whilst simultaneously reducing environmental emissions, by taking advantage of tax incentives. The motorist's choice of a particular fuel type may come to influence where and when a vehicle can be driven. However, the emergence of new options will not translate into genuine choice for all individuals. At the lower end of the income range, motoring is likely to become less affordable. In particular, the growing variation in the fuel efficiency of the vehicle fleet suggests that policies affecting the setting of motoring taxes are likely to remain controversial and contestable. As a result, tensions between government as revenue generator and government as promoter of traffic reduction and environmental protection policies can be expected to grow.

Whether new protests do emerge or not, it can be predicted that changes to the system of motoring taxation will sooner or later be sought by government. Arguments about the need to preserve fuel duty revenues as a necessity for maintaining public services were central to government rhetoric resisting protesters' demands for a large cut. The evidence presented above, though, suggests that, in any case, government can no longer depend on tax revenues from motoring continuing to increase. Indeed, they are likely to peak and decline in the next decade, particularly if the hypothesis that a ceiling of acceptability for duty rates has been reached is correct.

Rather than the peaking in motoring tax revenues being allowed to threaten dependent public services, or the balance of tax collection being shifted back towards direct sources, it seems much more likely that alternative motoring charges will be promoted as a more desirable option. For the first time, it may be that a policy for road pricing, although more technically complex, will come to look more straightforward in political terms than simply increasing fuel duties. Indeed, road pricing, already promoted by many policy actors for traffic restraint purposes, may emerge as the 'natural' solution for restoring lost tax revenues. Such a response can also be expected to restrain the traffic growth that is otherwise predicted to arise from the twin trends of higher incomes and lower per kilometre fuel costs. Depending on the level of tolls, part of the revenue could be hypothecated for public transport support, or for the maintenance and expansion of the trunk road network, or both.

If, as seems likely, these changes are combined with growing income inequality, motoring will become a more exclusive activity. If public transport fares also continue to rise, then current policy emerges as a recipe for greater social exclusion in transport. The issue of routine subsidies for public transport is hence likely to come back onto the political agenda; if not as a pro-environmental policy in itself, then as a means to make the application of such policies to car use acceptable.

If nothing is done to address growing inequality, then transport policy has the potential to again become an arena of conflict. Whether, in the final analysis, struggles and protests do emerge will depend on the ability of those individuals that are marginalized by transport policy to organise their individual grievances into a successful collective campaign, as the commercial interests managed to do in 2000.

References

Bolton, P. and Twigger, S. (2000). *Road Fuel Prices and Taxation.* UK Parliament House of Commons Library Research Paper 00/69. available at: http://www.parliament.uk/commons/lib/research/rp2000/rp00-069.pdf [accessed 01/10/01].

Bulgin, R. (1998). Toyota's marvellous mongrel. *The Daily Telegraph,* 21/3/98, motoring section.

Campbell, C.J. and Laherrier, J.L. (1998). The end of cheap oil? *Scientific American*, 278 (3), 78-83.

Commission of the European Communities (2000). *Commission starts research for super-economical car.* Press Release IP/00/1060, 26 September 2000, Brussels.

Commission for Integrated Transport (2000). *Pollution from Older Vehicles.* London, DETR.

Dargay, J. and Gately, D. (1997). Vehicle ownership to 2015: implications for energy use and emissions. *Energy Policy*, 25 (14-15), 1121-1127.

DETR (1997). *National Road Traffic Forecasts (Great Britain) 1997.* London, DETR.

DETR (1998). *A New Deal for Transport: Better for Everyone.* White Paper on Transport Policy. London, The Stationary Office.

DETR (2000a). *Tackling Congestion and Pollution: The Government's first report under the Road Traffic Reduction (National Targets) Act 1998.* London, DETR.

DETR (2000b). *Transport 2010: the 10 Year Plan.* London, DETR.

DETR (2000c). *Transport 2010: the Background Analysis.* London, DETR.

DETR (2000d). *Transport Statistics Great Britain 2000 Edition.* London, The Stationary Office.

DETR et al.. (2000). *The Air Quality Strategy for England, Scotland, Wales and Northern Ireland – Working Together for Clean Air.* CM 4548. London, The Stationary Office.

DETR (2001). *Traffic in Great Britain: 4th Quarter 2000.* Statistics bulletin (01)5, London, DETR.

Daily Telegraph (1999). The cat report. *The Daily Telegraph*, 9/11/99, motoring section.

Daily Telegraph (2001a). Used car prices rising. *The Daily Telegraph*, 16/6/01, motoring section.

Daily Telegraph (2001b). Car prices in UK edging up again. *The Daily Telegraph*, 26/7/01, motoring section.

English, A. (1998). Wunderkind rolls out with little to learn. *The Daily Telegraph*, 3/10/98, motoring section.

Foxall, J. (2001). Are you sitting comfortably? *The Daily Telegraph*, 05/05/01, motoring section.

HM Government (1998). *Road Traffic Reduction (National Targets) Act.* London, The Stationary Office.

HM Government (2000). *Transport Act.* London, The Stationary Office.

HM Treasury (2001). *Investing for the Long Term: Building Opportunity and Prosperity for All. National Budget for 2001-2.* London, HM Treasury.

Hudson, P. (2000). Brief drive: Toyota Prius. *The Daily Telegraph*, 19/12/00, motoring section.

Hudson, P. (2001). Brief drive: Honda Insight. *The Daily Telegraph*, 30/1/01, motoring section.

Hutton, W. (2000). Can Brown trump 'dump the pump'? *The Observer*, 9/7/00.

Institute for Fiscal Studies (2001). *Living Standards under Labour*. London, IFS. Available at: http://www.ifs.org.uk/election/ebn4.pdf (accessed 01/10/01).

Kågeson, P. (2000). *The Drive for Less Fuel.* Publication T&E 00/1, Brussels, European Federation for Transport and Environment. Available at: http://www.t-e.nu/Publications/2000%20pubs/T&E00-1.pdf (accessed 01/10/01).

Lovins, A.B. and Lovins, L.H. (1995). Reinventing the wheels. *The Atlantic Monthly*, January, 75-86.

Magoon, L.B. (2000). *Are We Running Out of Oil?* Poster published on US Geological Survey website. Available at: http://geopubs.wr.usgs.gov/open-file/of00-320/ (accessed 01/10/01).

Office for National Statistics (2001). *Family Spending. A report of the 1999-2000 Family Expenditure Survey.* London, The Stationary Office.

Orasch, W. and Wirl, F. (1997). Technological efficiency and the demand for energy (road transport). *Energy Policy*, 25 (14-15), 1129-1136.

Parkhurst, G.P. (2001). *New Road Vehicle Technologies and the Future of Motoring Taxation.* ESRC Transport Studies Unit, University College London, Research report 2001/2. Available at: http://www.ucl.ac.uk/~ucetgpp/roadtech.htm (accessed 01/10/01).

Plowden, W. (1971). *The Motor Car and Politics 1896-1970.* London, Bodley Head.

Sakaguchi, T. (2000). Influence of diffusion of fuel-efficient motor vehicles on gasoline demand for individual user owned passenger cars. *Energy Policy* 28, 895-903.

TRI/ECU (2000). *Choosing Cleaner Cars: The Role of Labels and Guides.* TRI Record 00/10/02. Transport Research Institite, Napier University also for Environmental Change Unit, University of Oxford. Available at: http://www.eci.ox.ac.uk/eep/ChoosingCleanerCarsReport.pdf (accessed 01/10/01).

Van den Brink, R.M.M. and Van Wee, B. (2001). Why has car-fleet specific fuel consumption not shown any decrease since 1990? Quantitative analysis of Dutch passenger car-fleet specific fuel consumption. *Transportation Research Part D*, 6, 75-93.

Ward, L. (2001). Earnings inequality widens under Labour. *The Guardian*, 18/4/01.

Whiteing, A., Coyle, M. and Bamford, C. (2002). Effective Fuel Management in Road Transport Fleets. In Lyons, G. and Chatterjee, K. (eds.) *Transport Lessons from the Fuel Tax Protests of 2000*, Aldershot, Ashgate.

Zachariadis, T., Ntziachristos, L., and Samaras, Z. (2001). The effect of age and technological change on motor vehicle emissions. *Transportation Research Part D*, 6, 221-7.

14 What can be Learned by Studying Transport Crises?

PETER BONSALL

Introduction

This chapter, placed quite deliberately at the end of the book, seeks to explore the wider applicability of conclusions drawn from the observation of transport crises. It asks what might be learned to assist in the management of future crises or in the furtherance of policy objectives. More generally, it asks whether the behaviour of individuals, groups and systems during a crisis can help us to understand or predict behaviour in other situations. Particular attention will be paid to evidence from the fuel crisis in September 2000 but evidence from other transport crises will be referred to where appropriate.

For the purposes of this chapter, a transport crisis is taken to be a relatively short period of time during which key features of a transport system are seriously compromised. Classic examples thus include massive reductions in capacity due to structural failure, safety restrictions or severe weather; significant reductions in levels of service due to industrial action or technical failure; and shortages of fuel due to interruption of the supply chain.

Such crises offer an opportunity to study aspects of behaviour or system performance under unusual conditions and so perhaps gain insights which might assist our understanding of the fundamental factors affecting that behaviour or performance. In scientific terms we can be said to be observing an 'experiment' which enables us to gather data on parts of a distribution which are rarely visible. These new data points might allow us to learn more about the shape of the overall distribution and so be better placed to predict its value in other situations. But are these observations anything more than outliers? Can observations made in such unusual circumstances reveal enough to provide a basis for prediction of behaviour or system performance in other circumstances?

During crises, people are forced to alter their behaviour in response to new constraints but may also be faced with new opportunities. The precise combination of constraints and opportunities may never be repeated but may nevertheless provide an insight into underlying preferences and decision-making processes. As noted by Thorpe et al. (2002), a crisis may be particularly helpful in the identification of factors which act as barriers to change and, depending on whether the change is desirable, might therefore provide pointers for the development of policy.

It is widely believed that, even if individual behaviour is difficult to predict, some aspects of the performance of transport systems can be represented quite effectively using relatively simple models. Thus for example, it is suggested that traffic speeds can be predicted as a function of capacity and flow, and that accidents and emissions can be estimated as a function of flow. If these simple relationships hold under crisis conditions then their transferability to other unobserved situations looks more assured - their domains of applicability can be said to have been expanded. But if the relationships do not appear to hold under crisis conditions, questions will need to be asked about the reason for this and, depending on the answers, some revision to the model may be required. In either event, investigation of the relationships under crisis conditions will have been helpful unless it has consumed resources which would otherwise have led to a greater improvement in our understanding of the performance of transport systems.

A key question is thus whether the study of transport crises is likely to yield new insights into the behaviour of individuals and systems or whether it is simply a distraction from the study of more relevant datasets.

This chapter begins with a summary of the 'findings' which have been identified by previous authors in this book and elsewhere. It then suggests reasons for caution when drawing conclusions from such findings and, in the light of these cautions, identifies some conclusions which can safely be drawn. It ends with a reflection on the title of the book by considering what lessons should be drawn from the fuel crisis of September 2000.

Key Findings

Findings Relating to the Fuel Crisis of 2000

Analysts and commentators on the UK fuel crisis in September 2000 have identified a variety of findings which might have wider application. Many of them are covered in greater detail elsewhere in this book but, for convenience, they are grouped here into three categories: those relating to the politics of the crisis; those relating to industry and services; and those relating to individuals' behaviour during the crisis.

In the *political arena* it has been noted (Goodwin, 2002) that the crisis developed very quickly and took the government, and most commentators, by surprise. The fuel tax escalator had been largely unnoticed during a period of falling world oil prices but quickly became an issue when world oil prices began to increase significantly. It appears that, despite warning signs during the summer, few people predicted the ability of the protesters to organise such an effective campaign or the depth of hostility among the general public to the then levels of fuel duty. The fuel blockaders tapped into a vein of support from the media and the general public because they were able to portray themselves as the champions of the motorist and tax payer fighting a tax-hungry and anti-car government (Rosen, 2002). The Government must have been concerned by the success of the protesters' action but were probably even more worried by the apparent support they received from the public. Clearly the Government were sufficiently worried to offer substantial tax cuts to hauliers in the months following the protest. This has led some commentators to conclude that further increases in fuel tax can no longer be on the agenda (Parkhurst, 2002). It may also have contributed to a retreat from several aspects of transport policy which might have been seen as anti-car.

The effects of the fuel shortage on *the economy* were dramatic. Almost every aspect of the national economy was seen to be dependent on road transport. The prevalence of Just-In-Time delivery practices resulted in the effects of interruptions in supply being felt very quickly. The complexity of supply networks caused some processes and services to be brought to a halt by interruptions to supplies or staff absences in some seemingly remote or trivial activity further up the chain. One case, publicised at the time, was of surgical operations halted by a shortage of hypodermic needles which was due to a shortage of high-grade packaging materials which in turn was due to staff absences at the plant manufacturing the packaging material. Most hauliers were able to continue deliveries - though with some reduction in levels of service. Although small independent hauliers without access to their own supplies suffered, the larger firms were not seriously inconvenienced by the shortage (Whiteing et al., 2002) and many have gone on to report very good profits for the year. Interestingly, by September 2001 the hauliers' Association was complaining of a shortage of drivers due to 'the misleading impression created during September 2000 that the industry was in crisis' (RHA, 2001). The events of September 2000 may have accelerated the trend towards domination of the industry by larger more profitable haulage companies.

In respect of *individuals' behavioural responses* it has been noted that an atmosphere of crisis, fanned by press and TV coverage, spread very quickly across the country. This led people to engage in typical supply-shortage activities. Panic-buying of fuel and foodstuffs was widespread and inevitably created the very shortages that people were seeking to avoid. Individual drivers devoted considerable efforts to seeking out available supplies – diverting from their normal routes, phoning round the local garages, or even setting out on hunting expeditions to look for fuel. Instances of ill-temper became increasingly common and caused commentators to lament the decline in the National Character. As in previous fuel shortages, there were reports of use of prohibited fuels and of fuel theft.

Despite the 'crisis' atmosphere, few people actually ran out of fuel (Chatterjee and Lyons, 2002; Marsden and Beecroft, 2002). Most people managed to eke out their reserves by changing their behaviour. 'The Fuel Crisis' became a socially acceptable excuse for cancellation of meetings and failure to attend meetings or to attend work. Trips were prioritised with less essential ones being curtailed in order to save fuel for the more important ones. Leisure trips and non-food shopping trips were usually the first to be cancelled while work and food shopping were the last (Chatterjee and Lyons, 2002; Thorpe et al., 2002). Trips to out-of town shopping centres and long distance trips on motorways were reduced while use of local shops increased. Efforts were made to increase fuel efficiency by adopting less aggressive driving styles (the observed *increases* in average speeds were not inconsistent with this – they resulted from the lower levels of congestion due to reduced volumes of traffic - Eves et al., 2002).

Changes in mode of transport were relatively uncommon although there was some increase in walking, cycling, car sharing and the use of public transport. Changes of mode were most common for commuting and school-escort trips (Chatterjee and Lyons, 2002). The use of rail and LRT increased more than that of bus with some evidence to suggest that some people who normally used bus used it *less* during the strike than at normal times. The use of telecommunications (particularly phone and email) increased - in some cases to substitute for physical travel but in others to explain absences and to seek information on fuel sources (the increased attention to TV news is also partly attributed to drivers seeking information about available supplies).

More General Findings Derived from Previous Transport Crises

Studies of previous transport crises have provided a wealth of data. Studies of the impact of industrial action on UK railways and London

Underground, of the collapse of the Tasman Bridge, and of earthquakes in San Francisco, Los Angeles and Kobe have been particularly fruitful sources. (See, variously, Lock and Gelling, 1976a,b; Deakin, 1991; Hanson and Weinstein, 1991; Wesemann et al., 1995; Kitamura et al., 1998). The following findings are commonly observed:

- When key links in a network are taken out, drivers will re-route and a new 'equilibrium' pattern generally emerges quite quickly.
- Peoples' initial behavioural responses may be maintained only for a short time. If the 'crisis' persists for longer than a few days, new patterns of behaviour will emerge as people assess the relative merits of the alternatives available to them.
- When a reduction in capacity increases peak period congestion, many people respond by peak spreading (particularly by travelling to work earlier).
- If a driver's 'normal' journey by car becomes impossible, their first preference is generally to re-route and their second is to change the time of travel. Changes of mode are not generally adopted and many drivers would sooner cancel the trip than attempt to make it by public transport.
- During crises, some trips are postponed or cancelled. Leisure trips are more likely to be suppressed than work, education or employers' business trips.
- When the supply of motor fuel is limited, drivers will prioritise their use of the car - cancelling some trips (or using alternative modes) in order to conserve stocks for more important ones.
- Pre-crisis travel patterns are usually restored quite soon after the end of a crisis.
- Although a crisis may directly affect the transport system, some of the changes to travel patterns are due to changes in activity patterns consequent on changes in the location of opportunities, rather than directly due to changes in the transport system.
- During crises, newspapers and television tend to exaggerate the crisis and so make it more serious.
- Even when the public initially support industrial action, this support may evaporate rapidly when the impacts become really serious.
- Although many people act selfishly (panic buying, hoarding, queue jumping) a crisis will often engender a mood of common purpose and lead to individual acts of kindness which would not be contemplated in 'normal' circumstances.

Reasons for Caution when Drawing Conclusions from Crises

It would be extraordinary if the experience of the fuel crisis of September 2000, and of other transport crises, were completely ignored by those who were affected by them. Government, protesters, hauliers, manufacturers, retailers and service providers will surely all have held debriefing sessions with the aim of being better prepared for the next crisis. Individuals will no doubt have formed a view on how to respond next time they are faced with a crisis. Consciously or otherwise, organisations and individuals will be applying knowledge gained during the crisis to non-crisis situations. Some analysts will be poised to recalibrate their models to reflect the new data. Some policy makers and pundits will be tempted to use behaviours and attitudes observed during the crisis to draw conclusions on future directions for transport policy.

But is this wise? Can any lasting or transferable lessons be learned from crises? Do they represent windows on underlying truths or are they relevant only to the circumstances of the crisis in which they were observed? Some reasons for *not* accepting them as transferable lessons are outlined below.

Behaviours observed during a crisis of short duration might not be sustainable over a longer period. For example, if the September 2000 fuel blockade had lasted longer:

- people might have found it impossible to continue to postpone certain categories of trip;
- public support for the blockaders might have evaporated once the effects of the fuel shortage became more serious and began to impinge on individual health or wealth;
- industries might not have continued to turn a blind eye to staff absences;
- vital industries and services would not have been able to continue to function; and
- government might have found it necessary to intervene more forcefully to bring the disruption to an end.

Behaviours adopted during an unexpected crisis will not have benefited from advance planning. If there had been advance notice of the September 2000 fuel shortage:

- more drivers might have discovered feasible public transport services;
- more drivers might have been able to arrange to share cars;

- more effective prioritisation and combination of trips might have been possible;
- individuals, firms and organisations would have increased their stocks of fuel and vital supplies;
- industry might have made arrangements to use telecommunications as a substitute for travel and might have made efforts to ensure cover for vital staff who were likely to find it difficult to get to work; and
- government might have succeeded in preventing the blockades or minimizing their impact.

Behaviour during a crisis generally represents a tactical *adaptation to the immediate situation rather than* strategic *response to a changed set of circumstances.* If it became clear that the disruption to supplies were likely to become permanent or semi-permanent, individuals, firms and organisations would no doubt consider more strategic responses. For example:

- individuals or firms might convert to alternative fuels or purchase vehicles with much greater fuel efficiency;
- individuals might move residence or job in order to be closer to public transport or in order to make walking or cycling a feasible option;
- public transport operators might extend their networks or increase their capacity or level of service;
- governments might intervene to ensure adequate provision of public transport;
- supply chains might be rationalised to reduce the distances that goods need to be moved; and
- retailers might seek locations well served by public transport.

Behaviours adopted during a crisis might not be regarded as acceptable during more normal times. For example:

- without the plausible excuse of a fuel crisis, absences from work, late arrival at appointments, cancellation of meetings, staying in the city over-night instead of returning home, refusal to chauffeur the children and similar indulgences might not be tolerated by employers, colleagues or spouses; and
- there is perhaps something in the national character which brings people together in a crisis and leads to cooperation and a desire to share in the common woe. This may have led to an increase in lift-giving and,

might even have tempted some drivers to indulge in that ultimate in shared pain – a journey by bus.

The attributes of options available during a crisis are not the same as those of the equivalent options at other times. For example, during the fuel crisis, public transport was more crowded whereas, when public transport is disrupted, roads tend to be busier. Decisions taking during a crisis may, or may not, have allowed for this. Initial impressions of new options experienced for the first time during a crisis will inevitably reflect their, usually somewhat degraded, performance during crisis conditions.

Some of the 'facts' collected during a crisis may be misleading. Perspectives on what happened during a crisis may suffer from selective recall and subjective reporting. For example:

- contemporaneous press reports are likely to reflect the 'line' being taken by that newspaper at the time (e.g. a desire to stoke up a crisis or attack the government);
- anecdotes and personal memories tend to exaggerate unusual events rather than the continuation of routine; and
- a voluntary survey may result in over-representation of people who feel they have something interesting to relate (which may mean an over-representation of people who were seriously affected by the crisis).

'Facts' may be open to misinterpretation. For example:

- initial reports are unlikely to have been corrected for seasonal effects - see for example Marsden and Beecroft (2002) on the importance of making such corrections when analysing listener/viewer statistics for TV, radio and cinema during the crisis;
- initial reports may attribute to the crisis things which may more properly be due to other, contemporaneous, circumstances - see for example Eves et al. (2002) on the importance of allowing for meteorological conditions when analysing air quality data;
- some of the behavioural change during a crisis may be attributable to week-on-week variation and 'churn' rather than to the crisis. This may, for example, explain some of the counter-intuitive changes in behaviour noted in the transition tables presented by Thorpe et al. (2002); and
- maintenance of past behaviour may simply reflect the fact that the person concerned was not affected by the crisis (because, for example, they had adequate fuel supplies).

Even if the circumstances of two crises are superficially similar, each one is a unique event. Consider, for example, the possible outcomes if...

- the fuel blockade had coincided with school holidays, or if
- the fuel blockade had coincided with bad weather, or if
- the new national public transport information service (Traveline) had been fully developed.

Political lessons are particularly likely to be very context-specific. Consider, for example:

- If the world price of fuel had not increased so dramatically over the summer of 2000...would the tax issue have been so emotive?
- If the French Government had not capitulated in the face of protests from its haulage industry earlier in the year (Salini, 2002)...would the UK protesters have gone ahead?
- If the tanker drivers had not been employed by a company whose haulage interests might have given it a reason to want to help pressurise government to reduce fuel tax ...would the blockade have been so effective?
- If the media had not supported the blockaders as champions of individual liberty...would the public have accepted the disruption so willingly?
- If the Government had not been due to hold a general election within the next year...would the Government response to the protests have been different?
- And if John Prescott had not owned two Jaguars and driven the few hundred yards to the party conference shortly before the fuel crisis erupted...would the Government have been able to press the environmental case for high fuel taxes?

The very occurrence of a crisis will have changed the world forever! This rather dramatic statement refers to the fact that the experiences gained during a crisis are bound to affect the behaviour of individuals and organisations in future crisis, and non-crisis, situations and will therefore complicate any attempt to use observation of behaviour during a crisis to forecast future behaviour. If another blockade of fuel supplies were to occur relatively soon after the last one, the blockaders and government would doubtless refine their tactics while hauliers, retailers, industry and individual travellers would seek to avoid the problems they experienced last time. Even if the crisis did not recur, some people and organisations will have made strategic adjustments in order to reduce their exposure to the kind of problems experienced during the crisis. Others will have concluded that something they tried for the first time during the crisis was worth repeating even in non-crisis situations.

Conclusions that can be Drawn from the Fuel Crisis of September 2000

Given the reservations expressed above, what conclusions can safely be drawn? Clearly we should not expect behaviour observed during the fuel crisis of 2000 to provide a snapshot of behaviour at any time in the future. There is no reason to expect the level of mode changes, or any other changes, to be repeated. Nor, given the different factors affecting the different changes, is there any reason to expect the *relative* extents of the different changes ever to be repeated. However, some very important conclusions can and should be drawn from the fuel crisis. The following are perhaps among the most important.

A relatively small group of protesters was able to force the Government to give in on a major item of taxation and contributed to a Government retreat from an important area of transport policy. Their success was due to a well-targeted campaign, effective use of communications technology, the tacit support of the tanker drivers and of much of the press and broadcast media, and a decision to call off their action before it really began to hurt. *It would be surprising if this weapon were not used again.*

Many people assume that motorists are over-taxed. *Given the success of the protesters in September 2000, and unless/until the environmental case for high fuel taxes is effectively made, it is unlikely that an attempt to increase the total burden of tax on motorists would now be accepted without protest.*

Current business practices leave industry and services very exposed to disruption of its supply chains. Finely tuned production systems have very little slack and it is not easy to identify elements in the chains which, if interrupted, would not quite quickly jeopardise the overall system. Industry is well aware of this but calculates that the risks do not justify the expense of maintaining higher stock levels or alternative sources of supply. *Against this background it is no longer so easy to determine which industries or individuals should have priority access to scarce fuel supplies because, within a very short period of time, any one of a myriad of seemingly trivial elements in the supply chain could become critical to the maintenance of that chain.*

Transport is overwhelmingly dependent on one type of fuel. *This leaves the economy dangerously vulnerable to disruptions in supply and to dramatic increases in price.*

Many people have, through their residential, employment and lifestyle decisions, become utterly dependent on the car. The evolution of dispersed land-use patterns and the loss of local facilities, particularly in rural areas, has contributed to a situation where the car has become a necessity for most people. Public transport is simply not a viable option for many journeys. *The fuel crisis has served to emphasise the radical change that would be necessary in land-use patterns, the provision of public transport, the structure of the economy and in individual lifestyles if car use is to be reduced.*

Even where there was a public transport alternative, many drivers were only vaguely aware of what it offered or how to access it. Many people did not know how to access information about public transport services – particularly buses. *The fuel crisis might have been less traumatic if people had already had some experience of finding out about, and preferably using, alternative modes of transport.*

Uncertainty about the options available during the crisis, whether it was the location of garages selling fuel, the availability of emergency parking in the city centre, or the time of the next bus to the city, was a major source of frustration. *Access to up-to-date information is a major problem during transport crises.*

Transport crises can engender a sense of community and common purpose but the majority response is self-interest. *Panic buying and hoarding seem to be inevitable consequences of a perceived shortage.*

During crises people and organisations are forced to research options and adopt behaviours which they would not contemplate in more normal circumstances. Some of these options and behaviours will turn out to be applicable to non-crisis situations. *Crises can act as catalysts to change.*

Behavioural responses during the fuel crisis cannot be assumed to be directly indicative of the responses that might occur over the longer term. It is, however, worth noting the following facts:

- *most drivers were willing, during the fuel shortage, to reduce fuel consumption by driving less aggressively and by postponing or cancelling some journeys;*
- *many drivers, particularly in urban areas, were aware that their routine journeys might be made by an alternative mode;*
- *many drivers believed that, with better planning, they could reduce their car use; and*
- *nevertheless, even in 'crisis' conditions, most drivers were reluctant to change modes or, more generally, to adopt options with which they were not yet familiar.*

Conclusions on the Success of Attempts to Study the Fuel Crisis of September 2000

It is clear that a rich and varied dataset has been amassed through a combination of special surveys, use of secondary sources and collation of data that had been collected routinely. The value of secondary sources is particularly apparent in Marsden and Beecroft's (2002) attempt to put the crisis in context. The potential value of data which were being collected as part of routine monitoring is exemplified by Eves et al.'*s* (2002) analysis of traffic data during, and either side of, the crisis week. But, as an example of the value of speculative investment in research effort and the ability of determined researchers to mount a survey in record time with limited resources, one cannot fail to be impressed by the achievements of Lyons and his team (Chatterjee and Lyons, 2002: Lyons and Beecroft, 2002).

Of course there are some gaps in the data collected (e.g. the analysis of responses to the fuel shortage would have benefited from information about the acceptability of flexi-time and home-working at the individual's workplace) but, given the short planning horizon, the gaps are remarkably few. The use of open-ended questions in some of the surveys contributed to the richness of the dataset even though, as anticipated, it made analysis much more difficult. Several of the surveys (most notably that conducted by Beatty et al., 2002) included questions about the process of learning and change (for example; Did people experiment with new options they had not tried previously? How did the options compare with prior expectations? Did they expect to continue with the new behaviour patterns after the crisis?). This data is of particular interest but would have benefited from a follow-up survey to see if the new patterns were maintained. In fact, of course, the severe disruption to transport caused by the floods in Autumn 2000, and following the Hatfield rail crash, would have complicated the analysis of data from any such follow-up surveys.

Although there may have been other abortive or unsuccessful attempts to study events during the fuel crisis, a particular feature of the surveys and analyses reported in this volume is their successful completion at relatively low cost. This is attributable to the research teams' decision to short-cut the procedures normally involved in setting up a survey and to the sponsors' preparedness to act quickly. The overall success of the exercise was also due to the willingness of respondents to provide the information requested of them. It seems that, at least in respect of data collection, a crisis brings out the best in people!

The Lessons

The observation of transport crises can yield fascinating insights into the structure of industry, the nature of politics, the role of the media, and the dynamics of public opinion. It can also reveal much about the behaviour of individuals, and the performance of transport systems, under stress. Crises act as catalysts for change. Experiences during a crisis can be very persuasive to individuals in planning their futures and for governments in considering their policy agenda. The very fact that particular attitudes or behaviours or system performances were manifest, albeit perhaps only briefly and in unique circumstances, can be of enormous importance.

There are strong reasons for supposing that behaviour during a brief crisis is a poor predictor of behaviour during a more extended period with apparently similar constraints but the fuel crisis of September 2000 has undoubtedly provided evidence which allows a number of very useful lessons to be learned. Many of these lessons relate to the avoidance, management or amelioration of future crises and have been referred to in previous sections of this chapter. Others relate more generally to future transport and land-use policy and to the analysis of traveller behaviour.

Lessons for Policy

The main lessons for policy come perhaps from realising the extent to which the economy has become dependent on fuel and individuals have become dependent on the car. Encouragement of the adoption of alternative fuel sources and/or of dual-fuel vehicles may be seen as a strategic imperative.

The fuel crisis will have served to emphasise how ambitious it is to seek to reduce car dependency but it has also provided some useful pointers as to how it might be done. Buses, in their current guise, were not seen as a viable alternative to the car for very many journeys. A particular criticism of buses was that they did not go where the traveller wanted to go at the time they wanted to go there. This might suggest a renewed effort to promote development of demand-responsive services, or of car-clubs, to supplement more conventional bus services. The transition from car to rail or LRT seemed easier than the transition from car to bus. It may be that buses need to be given more rail-like attributes (e.g. greater priority over other traffic and more visible routemarkers).

Anecdotal evidence suggests that one reason for drivers' reluctance to use public transport and other alternative modes during the crisis was their lack of knowledge of how to do so. This confirms the need for improved information about alternative modes and for the encouragement of experimentation with alternatives – for example by providing free 'taster' tickets and personalised advice. Surveys conducted during the crisis suggested that people thought that, with better planning, they could reduce their car use – this also suggests a role for personalised advice.

It is worth remarking that most of the behavioural adaptations observed during the fuel crisis were wholly in line with current policy objectives (reduced car-use, increased use of sustainable modes, increased use of local services...and so on). This points to the value of policy levers which might encourage people to treat fuel as a more valuable resource. Further increasing the tax on fuel might seem a bizarre recommendation following a crisis caused by protests at the existing levels of fuel tax, but that does not mean that it is not the right conclusion! Recognising the political sensitivity of any increase in fuel tax, it would be necessary to be seen to be simultaneously reducing other taxes on the road-user and providing relief for rural areas. Thus the fuel tax increase might be accompanied by abolition of Vehicle Excise Duty, funding third party insurance from the Exchequer and providing subsidies for rural post offices and other local services.

Lessons for Analysts

The student of human nature can learn much during crises - not least about the cusp which separates the friendly cooperation between people unaccustomed to using public transport from the determination of the person in the petrol queue to prevent anyone from getting ahead of him.

The traffic analyst can use observations of the relationship between traffic flow and speed during the petrol crisis to confirm the need to consider behavioural variables (in this case desired speeds) when attempting to forecast network speeds.

The analyst of individual traveller behaviour can find evidence to show that people are forced to reassess their behaviour patterns during crises and, given appropriate data, can learn something about the formation of choice sets, information searching and learning strategies.

Although some insights into these issues may emerge from study of the fuel crisis of September 2000 much work remains to be done. The analysis of the fuel crisis of September 2000 has helped to highlight issues and to identify gaps in our knowledge and data. We analysts should learn what we can from the data that was collected during the fuel crisis but should also learn from the experience of the data collection exercise itself. We should treat it as a pilot and endeavour to have clearer research goals, more refined survey tools and more reliable procedures in place before the next opportunity is upon us. Given that transport 'crises' are fairly frequent events, it is perhaps surprising that we appear not to have learned this lesson before now.

And Finally...

Did the fuel crisis of September 2000 provide any unique insights into behaviour? Did it allow us a glimpse of otherwise unobservable truths and does the analysis of crises represent a cost-effective use of resources? There can be little doubt that the observation and analysis of crises is crucial to the better management and amelioration of future crises but can it provide useful lessons for 'normal' situations?

Given the option of a carefully planned, controlled and monitored experiment lasting months or years it is unlikely that the observation of a crisis would add much. But in fact, of course, that option is unlikely ever to be available. The resources and influence required to provide full observation and experimental control are inconceivable in normal circumstances. The experimenter would never be allowed to interrupt fuel supplies or otherwise severely restrict peoples' lifestyles without their consent (noting of course that a voluntary restriction is no substitute, the absence of consent is a key feature of a forced change). As an 'experiment' the fuel crisis left much to be desired – but it did at least happen!

The fuel crisis of September 2000 stimulated an investment of intellectual effort and resources which has yielded a very useful body of data and some very valuable analyses. It served to concentrate attention and resources on important issues and to take forward our thinking in these areas further and faster than would otherwise have been the case. Although many of the 'lessons' are not new, they will have been strengthened by the additional evidence presented and discussed in this book.

References

Beatty S., Meadows M. and White D. (2002). Stages of Change in Drivers' Willingness to Reduce Car Use, Before and During the UK Fuel Crisis. In Lyons, G. and Chatterjee, K. (eds.) *Transport Lessons from the Fuel Tax Protests of 2000*, Aldershot, Ashgate.

Chatterjee, K. and Lyons, G. (2002). Travel Behaviour of Car Users During the UK Fuel Crisis and Insights into Car Dependence. In Lyons, G. and Chatterjee, K. (eds.) *Transport Lessons from the Fuel Tax Protests of 2000*, Aldershot, Ashgate.

Deakin, E. (1991). Transportation impacts of the 1989 Loma Prieta Earthquake: The Bay Bridge Closure, University of California Transportation Centre, working paper UCTC 294.

Eves D., Quick J., Boulter P. and Hickman, J. (2002). The Effect of the Fuel Crisis on Sections of the English Motorway Network. In Lyons, G. and Chatterjee, K. (eds.) *Transport Lessons from the Fuel Tax Protests of 2000*, Aldershot, Ashgate.

Goodwin, P. (2002). Are Fuel Prices Important? In Lyons, G. and Chatterjee, K. (eds.) *Transport Lessons from the Fuel Tax Protests of 2000*, Aldershot, Ashgate.

Hanson, M. and Weinstein, S (1991). Studies on the Loma Prieta Earthquake No. 5: East Bay Ferry Service, University of California Transportation Centre, working paper UCTC 162.

Kitamura, R., Yamamoto, T. and Fujii, S. (1998). Where did all the Traffic Go? Impacts of the Hanshin-Awaji Earthquake on Traffic and Travel. Annex to Cairns S, Hass-Klau C and Goodwin P (1998) *Traffic Impact of Highway Capacity Reductions: Assessment of the Evidence*, London, Landor Publishing.

Lock, J.B. and Gelling, M.J. (1976a). The Tasman Bridge Disaster Before and After, *Australian Road Research*, 6 (2), June, 9-16.

Lock, J.B. and Gelling, M.J. (1976b). Effect on Traffic Patterns of the Collapse of the Tasman Bridge, *Proceedings of Australian Road Research Board*, 8, 42-52.

Lyons, G. and Beecroft, M. (2002). Experiences and Lessons From Conducting an Emergency Survey of Motorists During the UK Fuel Crisis. In Lyons, G. and Chatterjee, K. (eds.) *Transport Lessons from the Fuel Tax Protests of 2000*, Aldershot, Ashgate.

Marsden, G. and Beecroft, M. (2002). Crisis or Drama? - A Summary of the Response to the Fuel Crisis in the UK. In Lyons, G. and Chatterjee, K. (eds.) *Transport Lessons from the Fuel Tax Protests of 2000*, Aldershot, Ashgate.

Parkhurst, G. (2002). The Top of the Escalator? In Lyons, G. and Chatterjee, K. (eds.) *Transport Lessons from the Fuel Tax Protests of 2000*, Aldershot, Ashgate.

RHA, (2001) Interview with Road Haulage Association spokesman broadcast on BBC Radio 4's "Today", 10th Sept 2001.

Rosen, P. (2002). Pro-Car or Anti-Car? 'Environment', 'Economy' and 'Liberty' in UK Transport Debates. In Lyons, G. and Chatterjee, K. (eds.) *Transport Lessons from the Fuel Tax Protests of 2000*, Aldershot, Ashgate.

Salini, P. (2002). Fuel Taxation and an Account of the Fuel Tax Protests in France. In Lyons, G. and Chatterjee, K. (eds.) *Transport Lessons from the Fuel Tax Protests of 2000*, Aldershot, Ashgate.

Thorpe N., Bell M., Polak J. and Noland R. (2002). A Telephone Survey of Stated Travel Responses to Fuel Shortages. In Lyons, G. and Chatterjee, K. (eds.) *Transport Lessons from the Fuel Tax Protests of 2000*, Aldershot, Ashgate.

Wesemann, L. Hamilton, T. and Tabaie, S. (1995). Variations in Traveller Response to Damaged Freeways and Transportation System Changes Associated with the Northridge Earthquake, paper presented at Transportation Research Board 75th annual Meeting, Washington DC, January.

Whiteing, A., Coyle, M. and Bamford, C. (2002). Effective Fuel Management in Road Transport Fleets. In Lyons, G. and Chatterjee, K. (eds.) *Transport Lessons from the Fuel Tax Protests of 2000*, Aldershot, Ashgate.